THE INTERNET OF THINGS

THE INTERNET OF THINGS

KEY APPLICATIONS AND PROTOCOLS

Olivier Hersent
Actility, France

David Boswarthick
ETSI, France

Omar Elloumi
Alcatel-Lucent, France

A John Wiley & Sons, Ltd., Publication

Library of Congress Cataloging-in-Publication Data

Hersent, Olivier.
  The internet of things : key applications and protocols / Olivier Hersent,
David Boswarthick, Omar Elloumi.
    p. cm.
  Includes bibliographical references and index.
  ISBN 978-1-119-99435-0 (hardback)
  1. Intelligent buildings.   2. Smart power grids.   3. Sensor networks.   I. Boswarthick, David.
II. Elloumi, Omar.   III. Title.
  TH6012.H47 2012
  681′.2–dc23
                                                                        2011037210

A catalogue record for this book is available from the British Library.

ISBN: 9781119994350 (H/B)

Typeset in 10.5/13pt Times by Aptara Inc., New Delhi, India

Contents

**Part II   LEGACY M2M PROTOCOLS FOR SENSOR NETWORKS,
              BUILDING AUTOMATION AND HOME AUTOMATION**

List of Acronyms

6LoWPAN	6LoWPAN is the acronym of IPv6 over Low power Wireless Personal Area Networks and the name of a working group in IETF
ACL	Access Control List
ACSE	Association Control Service Element
AER	All Electric Range
AFE	Analog Front End
AIB	Application Layer Information Base
AIS	Application Interworking Specification
AMI	Automatic Metering Infrastructure
ANSI	American National Standards Institute
AODV	Advanced Ad-Hoc On-Demand Distance Vectoring
AP	Application Process
APDU	Application Protocol Data Unit
API	Application Programming Interface
aPoC	Application Point of Contact
APS	Application Support Sublayer
APSDE-SAP	Application Support Sublayer Data Entity Service Access Point
APSME-SAP	Application Support Sublayer Management Entity Service Access Point
APSSE-SAP	Application Support Sublayer Security Entity Service Access Point
ARIB	Association of Radio Industries and Businesses is a standardization organization in Japan
ASDU	Aps Service Data Unit
ASK	Amplitude-Shift Keying
BbC	KNX Backbone Controller
BCI	Batibus Club International
BEV	Battery Electric Vehicle
BO	Beacon Order
BPSK	Binary Phase Shift Keying
BTT	Broadcast Transaction Table

CAN	Controller Area Network
CAP	Contention Access Period
CBC MAC	CBC Message Authentication Code
CC	Consistency Check
CCA	Clear Channel Assessment
CCM*	Extension of Counter with CBC-MAC Mode of Operation
CD range	Charge Depleting Range
CENELEC	European Committee for Electrotechnical Standardization
CER	Communication Error Rate
CFP	Contention Free Period
CI	Control Information
CNF	M-Bus CONFIRM Message
CRC	Cyclical Redundancy Check
CRL	X.509 Certificate Revocation List
CRUD	Create, Read, Update, Delete
CS mode	Charge Sustaining Mode
CSL	Coordinated Sampled Listening
CSMA	Carrier-Sense, Multiple Access
CSMA/CA	Carrier-Sense Multiple Access with Collision Avoidance
CSMA/CD	Carrier-Sense Multiple Access with Collision Detection
D device	ETSI M2M device without local M2M capabilities and interfaced to a gateway via the mId interface
D' device	ETSI M2M device implementing ETSI M2M capabilities and the mId interface to the network domain (does not interface via a gateway)
DA	Device Application
DAG	Direct Acyclic Graph
DAG root	A Node within the DAG that has no outgoing edge
DAO	Destination Advertisement Object
DER	Distinguished Encoding Rule
dIa	ETSI M2M Reference point between an application and ETSI M2M service capabilities
DIB	Data Information Block
DIO	DODAG Information Object
DIS	DODAG Information Solicitation
DLL	Data Link Layer the layer 2 specified in the seven-layer OSI model
DLMS	Device Language Message Specification is a specification for Data exchange for meter reading, tariff and load control
DODAG	Oriented Direct Acyclic Graph
DODAG Version	Specific iteration ("Version") of a DODAG with a given DODAGID
DODAGID	The identifier of a DODAG Root
DR	Demand Response

DRH	Data Record Header
DSSS	Direct Spread Spectrum Destination
DTSN	Destination Advertisement Trigger Sequence Number
ED	Energy Detection
EFF	Extended Frame Format
EHS	European Home System
EIB	European Installation Bus
EIBA	The European Installation Bus Association
EMC	Electromagnetic Compatibility
EMS	Energy Management System
EN 50065-1	CENELEC standard for Powerline transmission on low-voltage electrical installations in the frequency range 3 to 148,5 kHz
EP	Enforcement Point
EPID	Extended PAN ID
ESI	Energy Services Interface
ESP	Energy Service Portal
eTag	Entity Tag
ETSI	European Telecommunications Standards Institute is an independent, nonprofit, standardization organization in the telecommunications industry
ETSI PLT	The ETSI Powerline working group
EUI	Extended Unique Identifier
EV	Electric Vehicle
EVCC	Electric Vehicle Communication Controller
EVSE	Electric Vehicle Charging Equipment
EXI	Efficient XML Interchange Encoding
FCC	Federal Communications Commission
FFD	Full Function Device
FHSS	Frequency Hopping Spread Spectrum
FLiRS	Frequently Listening Routing Slave
FSK	Frequency-shift keying is a frequency modulation scheme in which digital information is transmitted through discrete frequency changes of a carrier wave
GA	Gateway Application
GBA	Generic Bootstrapping Architecture
GCM	Galois/Counter Mode
GMO	Gateway Management Object
GO	Group Object
GRE	Gestionnaire de réseau de transport
GRIP	Gateway Remote Interface Protocol
HC	Header Compression
HEV	Hybrid Electric Vehicle

HLS	High-Level Security
HomePlug Alliance	The HomePlug Alliance is a group of electronics manufacturers, service providers, and retailers that establishes standards for power line communication
IANA	Internet Assigned Number Authority
I-Band	Industrial Band, see ISM
IC	Interface Class
IEC TC13	International Electrotechnical Commission, Technical Committee 13
IEEE	The Institute of Electrical and Electronics Engineers
IEEE 1901	IEEE 1901 is an IEEE working group developing a global standard for high speed Powerline communications
IEEE 802.15.4	IEEE 802.15.4-2006 is a standard that specifies the physical layer and media access control for low-rate wireless personal area networks
IEEE P1901.2	IEEE 1901.2 is an IEEE working group developing a Powerline communications standard for metering applications
IETF	Internet Engineering Task Force
IHD	In Home Display
IID	Interface Id
IO	Interface Object
IPHA	IP Host Application
IPHC	IP Header Compression
IPSO	Internet Protocol for Smart Objects is a industry alliance promoting Internet of Objects
ISM	Industrial Scientific and Medical
ISO	International Organization for Standardization
ISP	Intersystem Protocol
ITS	Intelligent Transport System
ITU	International Telecommunication Union is the specialized agency of the United Nations which is responsible for information and communication technologies
ITU G.9972	ITU G.9972 (also known as G.cx) is a recommendation developed by ITU-T that specifies a coexistence mechanism for networking transceivers
ITU G.hn	G.hn is the common name for ITU recommendation G.9960, a home network technology standard being developed under the International Telecommunication Union
ITU G.hnem	An ITU project addressing the home networking aspects of energy management
LC	Line Coupler
LDN	Logical Device Name

LLC	Logical Link Control layer
LLN	Low Bitrate and Lossy Network
LLS	Low-Level Security
LN	Logical Name
LonWorks	LonWorks is a networking platform created to control applications The platform is built on a protocol created by Echelon Corporation
LowPAN	Low-power Wireless Personal Area Networks
LQI	Link Quality Information
LRWBS	Low Rate Wide Band Services are emerging services on Powerline transmitting in the 2–4 MHz band
LV-MV	Low Voltage (less than 600 Volts) and Medium Voltage (in the order of magnitude of 20 000 Volts)
M2M	Machine-to-Machine
MAC	Media Access Control
MAS	M2M Authentication Server
MCPS	MAC Common Part Sublayer
MCPS-SAP	MAC Common Part Service Access Point
MDU	Multidwelling Unit
mIa	Reference Point between a M2M application and the M2M Service Capabilities in the Networks and Applications Domain
MIC	Message Integrity Protection Code
mId	Reference point between an M2M Device or M2M Gateway and the M2M Service Capabilities in the Network and Applications Domain
MLDE	MAC Layer Management Entity
MLME-SAP	MAC Layer Management Entity Service Access Point
MP2P	Multipoint To Point Traffic
MSBF	M2M Service Bootstrap Function
MSP	Manufacturer Specific Profile
MTU	Maximum Transmission Unit
NA	Network Application
NAN	Neighborhood Area Network
NAPT	Network Address and Port Translation
NIB	Network Information Base
NIF	Node Information Frame
NIP	Network Interworking Proxy
NIST	National Institute of Standards and Technology is a measurement standards laboratory in USA
NLDE-SAP	Network Layer Data Entity Service Access Point
NLME	Network Layer Management Entity
NLME-SAP	Network Layer Management Entity Service Access Point
NLSE-SAP	Network Layer Security Entity Service Access Point

NREL	National Renewable Energy Laboratory
NRZ	Nonreturn to Zero
NUD	Neighbor Unreachability Detection
OBIS	Object Identification System
OCP	Objective Code Point
OF	Objective Function
OFDM	Orthogonal Frequency-Division Multiplexing
OOK	On-off keying the simplest form of modulation that represents digital data as the presence or absence of a carrier wave
O-QPSK	Offset-Quadrature Phase-Shift Keying
OSI	Open Systems Interconnections
OTA	Over-the-Air
OUI	Organizationally Unique Identifier
P2MP	Point to Multipoint Traffic
PAA	PANA Authentication Agent
PaC	PANA Client
PAN	Personal Area Network
PAN ID	Personal Area Network Identifier
PANA	Protocol for Carrying Authentication for Network Access
PCT	Programmable Communicating Thermostat
PEV	Plug-in Electric Vehicle
PHEV	Plug-in Hybrid Electric Vehicle
PHR	Physical Header
PHY	Physical Layer
PIB	PAN Information Base
PIO	Prefix Information Option
PLC	Powerline Communication
PLT	Powerline Technology
PN	Parent Node
PoC	Point of Contact
PRE	PANA Relay Element
PRIME	Powerline Intelligent Metering Evolution
PSDU	Physical Service Data Unit
PSEM	Protocol Specification for Electric Metering
PSSS	Parallel Spread Spectrum modulation
PWM	Pulse Width Modulation
Rank	A node's individual position relative to other nodes with respect to a DODAG root
REQ	M-Bus REQUEST message
REST	Representational State Transfer
RFD	Reduced Function Device
RIT	Receiver-Initiated Transmission

ROLL	Routing over Low-power and Lossy network
RPF	Reverse Power Flow
RPL	RPL IPv6 Routing Protocol over Low-power and Lossy Networks
RPL Instance	A set of one or more DODAGs that share a RPLInstanceID
RPLInstanceID	A unique identifier within a RPL LLN. DODAGs with the same RPLInstanceID share the same Objective Function
RSP	M-Bus RESPOND Message
RTE	Réseau Transport Electricité
RTU	Remote Terminal Unit
RZtime	Rendezvous Time
SA	Secure Association
SAP	Service Access Point
S-Band	Scientific Band, see ISM
SCDE	Secured Connection Protocol
SCL	Service Capability Layer
SCME	SCoP Management Entity
SCoP	SCoP Data Entity
SCPT	Standard Configuration Property Type
SCSS	SCoP Security Service
SDP	SECC Discovery Protocol
SDU	Service Data Unit
SECC	Supply Equipment Communication Controller
SFD	Start Frame Delimiter
SHR	Synchronous Header
SKKE	Symmetric-Key Key Exchange
SLAAC	IPv6 Stateless Address Autoconfiguration
SN	Short Name
SND	M-Bus SEND Message
SNVT	Standard Network Variable Type
SoC	System on Chip
SUN	Smart Utility Network
TDMA	Time division multiple access is a channel access method for shared medium networks
TL	Transport Layer
TLS	Transport Layer Security
ToU	Time of Use
TP1	KNX Twisted Pair Physical Media
TSCH	Time-Synchronized Channel Hopping
TSO	Transmission System Operator
UC	Upgrade Client
UID	Unique Node Identifier
U-NII	Unlicensed National Information Infrastructure

UNVT	User Network Variable Type
US	Upgrade Server
V2GTP	Vehicle to Grid Transfer Protocol
VIB	Value Information Block
VIF	Value Information field, see M-Bus
WADL	Web Application Description Language
xAE	Application Enablement M2M Service Capability
xBC	Compensation Broker M2M Service Capability
XCAP	Extensible Markup Language (XML) Configuration Access Protocol (RFC 4825)
xCS	Communication Selection M2M Service Capability
xHDR	History and Data Retention M2M Service Capability
xIP	Interworking Proxy M2M Service Capability
xRAR	Reachability, Addressing and Repository M2M Service Capability
xREM	Remote Entity Management M2M Service Capability
xSEC	Security M2M Service Capability
xTM	Transaction Management M2M Service Capability
xTOE	Telco Operator Exposure M2M Service Capability
ZBD	ZigBee Bridge Device
ZC	ZigBee Coordinator
ZCL	ZigBee Cluster Library
ZCP	ZigBee Compliant Platform
ZDO	ZigBee Device Object
ZDP	ZigBee Device Profile
ZED	ZigBee End Device
Zero-crossing	In alternating current, the zero-crossing is the instantaneous point at which there is no voltage present
ZGD	ZigBee Gateway Device
ZigBee Alliance	ZigBee Alliance is a group of companies that maintain and publish the ZigBee standard
ZIPT	ZigBee IP Tunneling Protocol
ZR	ZigBee Router
ZSE	ZigBee Smart Energy

Introduction

Innovation rarely comes where it is expected. Many governments have been spending billions to increase the Internet bandwidth available to end users ... only to discover that there are only a limited number of HD movies one can watch at a given time. In fact, there are also a limited number of human beings on Earth.

The Internet is about to bring us another ten years of surprises, as it morphs into the "Internet of Things" (IoT). Your mobile phone and your PC are already connected to the Internet, maybe even your car GPS too. In the coming years your car, office, house and all the appliances it contains, including your electricity, gas and water meters, street lights, sprinklers, bathroom scales, tensiometers and even walls[1] will be connected to the IoT. Tomorrow, several improvements will be made to these appliances such as not heating your house if hot weather is forecast, watering your garden automatically only if it doesn't rain, getting assistance immediately on the road, and so on. These improvements will facilitate our lives and utilize natural resources more efficiently.

Why is this happening now? As always, there is a combination of small innovations that, together, have reached a critical mass:

- Fieldbus technologies, using proprietary protocols and standards (LON, KNX, DALI, CAN, ModBus, M-Bus, ZigBee, Zwave ...), have explored many vertical domains. Gradually, these domains have started to overlap as use cases expanded to more complex situations, and protocols have emerged to facilitate interoperability (e.g., BACnet). But in many ways, current fieldbus deployments continue to use parallel networks that do not collaborate. The need for a common networking technology that would run over any physical layer, like IP, has become very clear.
- Despite the need for a layer 2 independent networking technology for fieldbuses, IP was not considered as a possible candidate for low-bitrate physical layers typically used in fieldbus networks, due to its large overheads. But the wait is now over: with 6LoWPAN not only has IP technology found its way onto low-bitrate networks but – surprise, surprise – it is IPv6 ! As an additional bonus, the technology comes with a state-of-the-art, standardized IP level mesh networking protocol, which makes multiphy

---

[1] Sensors for structural monitoring.

mesh networking a reality: finally different layer 2 fieldbus technologies can collaborate and form larger networks.

- Today, local fieldbus networks optimize the HVAC[2] regulation in your office and perhaps your home, with sophisticated algorithms. The energy-efficiency regulation for new building construction has created a need for even more sophisticated algorithms, like predictive regulation that takes into account weather forecasts or load shifting that incorporates the CO_2 content of electricity. In many automation sectors, the current state-of-the-art tool requires the local fieldbus to collaborate with hosted centralized applications and data sources. The technology required to enable this progressed in steps: oBix introduced the concept of a uniform (REST) interface to sensor networks, ETSI M2M added the management of security and additional improvements required in large-scale public networks.

The industry was only missing a really, really compelling business case to trigger the enormous amount of R&D that will be required to integrate all these technologies and build a bulletproof Internet of Things.

This business case is coming from the energy sector:

- The accelerated introduction of renewable-energy sources in the overall electricity production park brings an increasing degree of randomness to the traditionally deterministic supply side.
- In parallel, the mass introduction of rechargeable electric and hybrid vehicles is making the demand side more complex: EVs are roaming objects that will need to authenticate to the network, and will require admission control protocols.

The current credo of electricity operators "demand is unpredictable, and our expertise is to adapt production to demand", is about to be reversed into "production is unpredictable, and our expertise is to adapt demand to production".

As the rules of the game change, the key assets of an energy operator will no longer be the means of production, but the next-generation communication network and information system, which they still need to build entirely, creating an enormous market for mission-critical M2M technology. This dramatic change of how electricity will be distributed prefigures the more general evolution of the Internet towards the Internet of Things, where telecom operators and network-based application developers will have an increasing impact on our everyday lives, including the things that we touch and use.

This book targets an audience of engineers who are involved or want to get involved in large-scale automation and smart-grid projects and need to get a feel for the "big picture".

Many such projects will involve interfaces with existing systems. We included detailed overviews of many legacy fieldbus and automation technologies: BACnet, CAN, LON, M-Bus/wMBUS, ModBus, LON, KNX, ZigBee, Z-Wave, as well as C.12 and

---

[2] Heating, ventilation and air conditioning.

DLMS/COSEM metering standards. We also cover in detail two common fieldbus physical layers: 802.15.4 and PLC.

This book will not make you an expert on any of these technologies, but provides enough information to understand what each technology can or cannot do, and the fast-track descriptions should make it much easier to learn the details by yourself.

The future of fieldbus protocols is IP: we introduce 6LoWPAN and RPL, as well as the first automation protocol to have been explicitly designed for 6LoWPAN networks: ZigBee SE 2.0. We also provide an introduction to the emerging ETSI M2M standard, which is the much-awaited missing piece for service providers willing to provide a general-purpose public M2M infrastructure, shared by all applications.

I would like to thank Paul Bertrand, the inventor of the lowest-power PLC fieldbus technology to date (WPC) and designer of the first port of 6LoWPAN to PLC for accepting to write – guess what – the Powerline Communications chapter of this book. I am also grateful for the C.12 and DLMS chapters that were provided by Jean-Marc Ballot (Alcatel), and required a lot of documentation work.

Despite my efforts, there are probably quite a few errors remaining in the text, but there would have been many more without the help of the expert reviewers of this book: Cedric Chauvenet for 6LoWPAN/RPL, Mathieu Pouillot for ZigBee, Juan Perez (EPEX) for the smart-grid section, François Collet (Renault) for EV charging, Alexandre Ouimet-Storrs for his insights on energy trading, and the companies who provided internal documentation or reviews: Echelon for LON (with special thanks to Bob Dolin, Jeff Lund, Larry Colton and Mark Ossel), and Sigma Designs for Z-Wave. I am also grateful to Benoit Guennec and Baptiste Vial (Connected Object), who supplied me with the temperature and consumption profiles of their homes and shared their field experience with Z-Wave. Please let me know of remaining errors, so that we can improve the next edition of this book, at olivier.hersent@actility.com.

Gathering and reading the documentation for this book has been an amazing experience discovering new horizons and perspectives. I hope you will enjoy reading this book as much as I enjoyed writing it.

Olivier Hersent

Part One

M2M Area Network Physical Layers

1

IEEE 802.15.4

1.1   The IEEE 802 Committee Family of Protocols

The Institute of Electrical and Electronics Engineers (IEEE) committee 802 defines physical and data link technologies. The IEEE decomposes the OSI link layer into two sublayers:

- The media-access control (MAC) layer, sits immediately on top of the physical layer (PHY), and implements the methods used to access the network, typically the carrier-sense multiple access with collision detection (CSMA/CD) used by Ethernet and the carrier-sense multiple access with collision avoidance (CSMA/CA) used by IEEE wireless protocols.
- The logical link control layer (LLC), which formats the data frames sent over the communication channel through the MAC and PHY layers. IEEE 802.2 defines a frame format that is independent of the underlying MAC and PHY layers, and presents a uniform interface to the upper layers.

Since 1980, IEEE has defined many popular MAC and PHY standards (Figure 1.1 shows only the wireless standards), which all use 802.2 as the LLC layer.

  802.15.4 was defined by IEEE 802.15 task group 4/4b (http://ieee802.org/15/pub/TG4b.html). The standard was first published in 2003, then revised in 2006. The 2006 version introduces improved data rates for the 868 and 900 MHz physical layers (250 kbps, up from 20 and 40 kbps, respectively), and can be downloaded at no charge from the IEEE at http://standards.ieee.org/getieee802/download/802.15.4-2006.pdf

1.2   The Physical Layer

The design of 802.15.4 takes into account the spectrum allocation rules of the United States (FCC CFR 47), Canada (GL 36), Europe (ETSI EN 300 328-1, 328-2, 220-1) and

The Internet of Things: Key Applications and Protocols, First Edition.
Olivier Hersent, David Boswarthick and Omar Elloumi.
© 2012 John Wiley & Sons, Ltd. Published 2012 by John Wiley & Sons, Ltd.

MAC layer		BAND
802.11	WiFi	802.11, 802.11b, 802.11g, 802.11n : ISM 802.11a : U-NII
802.15.1	Bluetooth	ISM 2.4 GHz
802.15.4	ZigBee, SLowPAN	ISM 2.4 GHz worldwide ISM 902–928 MHz USA 868.3 MHz European countries 802.15.4a: 3.1–10.6 GHz
802.16	Wireless Metropolitan Access Networks Broadband Wireless Access (BWA) WiMax	802.16 : 10–66 GHz 802.16a: 2–11 GHz 802.16e: 2–11 GHz for fixed/2–6 GHz for mobile

Figure 1.1   IEEE-defined MAC layers.

Japan (ARIB STD T66). In the United States, the management and allocation of frequency bands is the responsibility of the Federal Communications Commission (FCC). The FCC has allocated frequencies for industrial scientific and medical (ISM) applications, which do not require a license for all stations emitting less than 1 W. In addition, for low-power applications, the FCC has allocated the Unlicensed National Information Infrastructure (U-NII) band. Figure 1.2 lists the frequencies and maximum transmission power for each band.

   IEEE 802.15.4 can use:
- The 2.4 GHz ISM band (S-band) worldwide, providing a data rate of 250 kbps (O-QPSK modulation) and 15 channels (numbered 11–26);
- The 902–928 MHz ISM band (I-band) in the US, providing a data rate of 40 kbps (BPSK modulation), 250 kbps (BPSK+O-QPSK or ASK modulation) or 250 kbps (ASK modulation) and ten channels (numbered 1–10)
- The 868–868.6 MHz frequency band in Europe, providing a data rate of 20 kbps (BPSK modulation), 100 kbps (BPSK+O-QPSK modulation) or 250 kbps (PSSS: BPSK+ASK

FCC band	Maximum transmit power	Frequencies
Industrial Band	<1W	902 MHz–928 MHz
Scientific Band	<1W	2.4 GHz–2.48 GHz
Medical Band	<1W	5.725 GHz–5.85 GHz
U-NII	<40 mW	5.15 GHz–5.25 GHz
	<200 mW	5.25 GHz–5.35 GHz
	<800 mW	5.725 GHz–5.82 GHz

Figure 1.2   FCC ISM and U-NII bands.

modulation), and a single channel (numbered 0 for BPSK or O-QPSK modulations, and 1 for ASK modulation).

In practice, most implementations today use the 2.4 GHz frequency band. This may change in the future as the IP500 alliance (www.ip500.de) is trying to promote applications on top of 6LoWPAN and 802.15.4 sub-GHz frequencies and 802.15.4g introduces more sub-GHz physical layer options. More recently, a new physical layer has been designed for ultrawide band (3.1 to 10.6 GHz).

Overview of O-QPSK Modulation at 2.4 GHz

The data to be transmitted is grouped in blocks of 4 bits. Each such block is mapped to one of 16 different *symbols*. The symbol is then converted to a 32-bit chip sequence (a pseudorandom sequence defined by 802.15.4 for each symbol). The even bits are transmitted by modulating the inphase (I) carrier, and the odd bits are transmitted by modulating the quadrature phase (Q) carrier (Figure 1.3). Each chip is modulated as a half-sine pulse. The transmitted chip rate is 2 Mchip/s, corresponding to a symbol rate 32 times slower, and a user data bitrate of 250 kbps. The sum of the I and Q signals is then transposed to the 2.4 GHz carrier frequency.

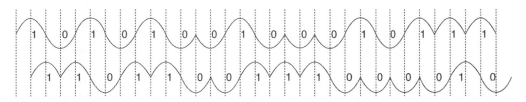

Figure 1.3   O-QPSK I and Q components.

802.15.4 uses a 32-bit encoding when it needs to refer to a specific frequency band, modulation, and channel. The first 5 bits encode a page number, and the remaining 27 bits are used as channel number flags within the page. The mapping of page and channel number to the frequency band, modulation and center frequency is shown in Figure 1.4.

1.2.1   Interferences with Other Technologies

Because the scientific band (2.4–2.48 GHz) is also unlicensed in most countries, this frequency band is used by many wireless networking standards, among which are WiFi (802.11, 802.11b, 802.11g, 802.11n), 802.15.4, and other devices such as cordless phones and microwave ovens.

Frequency band	Modulation	Page number	Channel number and center frequency
2.4 GHz	0-QPSK	0	11 : 2405 MHz
			12 : 2410 MHz
			13 : 2415 MHz
			14 : 2420 MHz
			15 : 2425 MHz
			16 : 2430 MHz
			17 : 2435 MHz
			18 : 2440 MHz
			19 : 2445 MHz
			20 : 2450 MHz
			21 : 2455 MHz
			22 : 2460 MHz
			23 : 2465 MHz
			24 : 2470 MHz
			25 : 2475 MHz
			26 : 2480 MHz
915 MHz	BPSK	0	1 : 906 MHz
	BPSK+ ASK	1	2 : 908 MHz
	BPSK+ 0–QPSK	2	3 : 910 MHz
			4 : 912 MHz
			5 : 914 MHz
			6 : 916 MHz
			7 : 918 MHz
			8 : 920 MHz
			9 : 922 MHz
			10 : 924 MHz
868 MHz	BPSK	0	0 : 868.3 MHz
	BPSK+ ASK	1	1 : 868.3 MHz
	BPSK+ 0–QPSK	2	0 : 868.3 MHz

Figure 1.4   802.15.4 frequency bands, modulations and channels.

1.2.1.1   FHSS Wireless Standards

The 802.11 physical layer uses frequency hopping spread spectrum (FHSS) and direct spread spectrum modulation. Bluetooth (802.15.1) uses FHSS in the ISM band.

The FHSS technology divides the ISM band into 79 channels of 1 MHz (Figure 1.5). The FCC requires that a transmitter should not use any channel more than 400 ms at a time (dwell time), and should try to use at least 75 channels (but this may not always be possible if some channels are too noisy).

FHSS Channel	Frequency (GHz)		
2	2.401–2.402		
3	2.402–2.403		
4	2.403–2.404		
...			
80	2.479–2.480		

Figure 1.5   FHSS channels defined by the FCC in the S-Band.

1.2.1.2   DSSS Wireless Standards

802.11b and 802.11g use only direct spread spectrum (DSSS). 11 DSSS channels have been defined, each of 16 MHz bandwidth, with center frequencies of adjacent channels separated by 5 MHz. Only 3 channels do not overlap (outlined in bold font in Figure 1.6): these channels should be used in order to minimize interference issues in adjacent deployments (3 channels are sufficient for a bidirectional deployment, however in tridimensional deployments, for example, in a building, more channels would be required).

1.2.2   Choice of a 802.15.4 Communication Channel, Energy Detection, Link Quality Information

In practice, only the 2.4 GHz frequency band is commonly used by the network and applications layers on top of 802.15.4, typically ZigBee and 6LoWPAN. The transmission power is adjustable from a minimum of 0.5 mW (specified in the 802.15.4 standard) to a maximum of 1 W (ISM band maximum). For obvious reasons, on links involving a battery-operated device, the transmission power should be minimized. A transmission power of 1 mW provides a theoretical outdoor range of about 300 m (100 m indoors).

DSSS channel	Frequency (GHz)
1	**2.404–(2.412)–2.420**
2	2.409–(2.417)–2.425
3	2.414–(2.422)–2.430
4	2.419–(2.427)–2.435
5	2.424–(2.432)–2.440
6	**2.429–(2.437)–2.445**
7	2.434–(2.442)–2.450
8	2.439–(2.447)–2.455
9	2.444–(2.452)–2.460
10	2.449–(2.457)–2.465
11	**2.456–(2.462)–2.470**

Figure 1.6   DSSS channels used by 802.11b.

Synchronous header (SHR)		Physical header (PHR)		Physical Service Data Unit
Preamble	SFD 111100101	Frame length (7 bits)	Ibit (reserved)	0 to 127 bytes

Figure 1.7 802.15.4 physical layer frame.

802.15.4 does not use frequency hopping (a technique that consumes much more energy), therefore the choice of the communication channel is important. Interference with FHSS technologies is only sporadic since the FHSS source never stays longer than 400 ms on a given frequency. In order to minimize interference with DSSS systems such as Wi-Fi (802.11b/g) set to operate on the three nonoverlapping channels 1, 6 and 11, it is usually recommended to operate 802.15.4 applications on channels 15, 20, 25 and 26 that fall between Wi-Fi channels 1, 6 and 11.

However, the 802.15.4 physical layer provides an energy detection (ED) feature that enables applications to request an assessment of each channel's energy level. Based on the results, a 802.15.4 network coordinator can make an optimal decision for the selection of a channel.

For each received packet, the 802.15.4 physical layer also provides link quality information (LQI) to the network and application layers (the calculation method for the LQI is proprietary and specific to each vendor). Based on this indication and the number of retransmissions and lost packets, transmitters may decide to use a higher transmission power, and some applications for example, ZigBee Pro provide mechanisms to dynamically change the 802.15.4 channel in case the selected one becomes too jammed, however, such a channel switch should remain exceptional.

1.2.3   Sending a Data Frame

802.15.4 uses carrier-sense multiple access with collision avoidance (CSMA/CA): prior to sending a data frame, higher layers are first required to ask the physical layer to performs a clear channel assessment (CCA). The exact meaning of "channel clear" is configurable: it can correspond to an energy threshold on the channel regardless of the modulation (mode 1), or detection of 802.15.4 modulation (mode 2) or a combination of both (energy above threshold *and* 802.15.4 modulation: mode 3).

After a random back-off period designed to avoid any synchronization of transmitters, the device checks that the channel is still free and transmits a data frame. Each frame is transmitted using a 30- to 40-bit preamble followed by a start frame delimiter (SFD), and a minimal physical layer header composed only of a 7 bits frame length (Figure 1.7).

1.3   The Media-Access Control Layer

802.15.4 distinguishes the part of the MAC layer responsible for data transfer (the MAC common part sublayer or MCPS), and the part responsible for management of the MAC layer itself (the Mac layer management entity or MLME).

The MLME contains the configuration and state parameters for the MAC layer, such as the 64-bit IEEE address and 16-bit short address for the node, how many times to retry accessing the network in case of a collision (typically 4 times, maximum 5 times), how long to wait for an acknowledgment (typically 54 symbol duration units, maximum 120), or how many times to resend a packet that has not been acknowledged (0–7).

1.3.1   802.15.4 Reduced Function and Full Function Devices, Coordinators, and the PAN Coordinator

802.15.4 networks are composed of several device types:

- 802.15.4 networks are setup by a *PAN coordinator* node, sometimes simply called the coordinator. There is a single PAN coordinator for each network identified by its PAN ID. The PAN coordinator is responsible for scanning the network and selecting the optimal RF channel, and for selecting the 16 bits PAN ID (personal area network identifier) for the network. Other 802.15.4 nodes must send an association request for this PAN ID to the PAN coordinator in order to become part of the 802.14.4 network.
- *Full Function Devices* (FFD), also called coordinators: these devices are capable of relaying messages to other FFDs, including the PAN coordinator. The first coordinator to send a beacon frame becomes the PAN coordinator, then devices join the PAN coordinator as their parent, and among those devices the FFDs also begin to transmit a periodic beacon (if the network uses the beacon-enabled access method, see below), or to respond to beacon requests. At this stage more devices may be able to join the network, using the PAN coordinator or any FFD as their parent.
- *Reduced Function Devices* (RFD) cannot route messages. Usually their receivers are switched off except during transmission. They can be attached to the network only as leaf nodes.

Two alternative topology models can be used within each network, each with its corresponding data-transfer method:

- The *star topology*: data transfers are possible only between the PAN coordinator and the devices.
- The *peer to peer* topology: data transfers can occur between any two devices. However, this is simple only in networks comprising only permanently listening devices. Peer to peer communication between devices that can enter sleep mode requires synchronization, which is not currently addressed by the 802.15.4 standard.

Each network, identified by its PAN ID, is called a *cluster*. A 802.15.4 network can be formed of multiple clusters (each having its own PAN ID) in a tree configuration: the root PAN coordinator instructs one of the FFD to become the coordinator of an adjacent PAN.

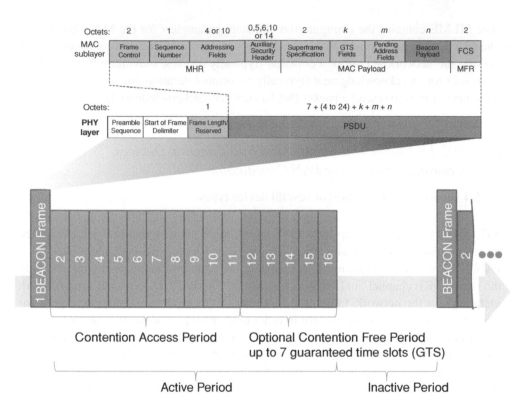

Figure 1.8   802.15.4 Superframe structure.

Each child PAN coordinator may also instruct a FFD to become a coordinator for another PAN, and so on.

The MAC layer specified by 802.15.4 defines two access control methods for the network:

- The *beacon-enabled access method* (or slotted CSMA/CA). When this mode is selected, the PAN coordinator periodically broadcasts a *superframe*, composed of a starting and ending beacon frame, 15 time slots, and an optional inactive period during which the coordinator may enter a low-power mode (Figure 1.8). The first time slots define the contention access period (CAP), during which the other nodes should attempt to transmit using CSMA/CA. The last N ($N \leq 7$) time slots form the optional contention free period (CFP), for use by nodes requiring deterministic network access or guaranteed bandwidth.

    The beacon frame starts by the general MAC layer frame control field (see Figures 1.8 and 1.9), then includes the source PAN ID, a list of addresses for which the coordinator has pending data, and provides superframe settings parameters. Devices willing to send data to a coordinator first listen to the superframe beacon, then synchronize to the

	Bytes	

Frame Control Field	2	000-------------: Beacon frame
		001-------------: Data Frame
		010-------------: Ack Frame
		011-------------: Command frame
		---1-------------: Security enabled at MAC layer
		----1-------------: Frame pending
		-----1-----------: Ack request
		------1----------: PAN ID compression
		     (source PAN ID omitted, same as destination)
		-------XXX------: reserved
		----------XX----:Destination address mode
		      *00 : PAN ID and destination not
		       present (indirect addressing)*
		      01 : reserved
		      10 : short 16-bit addresses
		      11 : extended 64-bit addresses
		--------- XX--:Frame version (00 : 2003, 01 : 2006)
		--------------XX:Source address mode

Sequence number	1	
Destination PAN ID	0 or 2	
Destination address	0 or 2 or 8	
Source PAN ID	0 or 2	
Source address	0 or 2 or 8	
Auxiliary security	variable	Contains security control, Frame counter, Key identifier fields
Payload	variable	
FCS	2	CRC 16 frame check sequence

Figure 1.9 802.15.4 MAC layer frame format.

superframe and transmit data either during the CAP using CSMA/CA, or during the CFP. Devices for which the coordinator has pending data should request it from the coordinator using a MAC data request command (see Figure 1.10).

When multiple coordinators transmit beacons, the active periods of the super frames should not overlap (a configuration parameter, *StartTime*, ensures that this is the case).

- The *nonbeacon-enabled access method* (unslotted CSMA/CA). This is the mode used by ZigBee and 6LoWPAN. All nodes access the network using CSMA/CA. The coordinator provides a beacon only when requested by a node, and sets the beaconorder (BO) parameter to 15 to indicate use of the nonbeacon-enabled access method. Nodes (including the coordinator) request a beacon during the *active scan* procedure, when

01	Association request
02	Association response
03	Disassociation notification
04	Data request
05	PAN ID conflict notification
06	Orphan notification
07	Beacon request
08	Coordinator realignment
09	GTS request

Figure 1.10   802.15.4 command identifiers.

trying to identify whether networks are located in the vicinity, and what is their PAN ID.

The devices have no means to know whether the coordinator has pending data for them, and the coordinator cannot simply send the data to devices that are not permanently listening and are not synchronized: therefore, devices should periodically (at an application defined rate), request data from the coordinator.

1.3.2   Association

A node joins the network by sending an association request to the coordinator's address. The association request specifies the PAN ID that the node wishes to join, and a set of capability flags encoded in one octet:

- *Alternate PAN:* 1 if the device has the capability to become a coordinator
- *Device type:* 1 for a full function device (FFD), that is, a device capable of becoming a full function device (e.g., it can perform active network scans).
- *Power source:* 1 if using mains power, 0 when using batteries.
- *Receiver on while transceiver is idle:* set to 1 if the device is always listening.
- *Security capability:* 1 if the device supports sending and receiving secure MAC frames.
- *Allocation address:* set to 1 if the device requests a short address from the coordinator.

In its response, the coordinator assigns a 16-bit short address to the device (or 0xFFFE as a special code meaning that the device can use its 64-bit IEEE MAC address), or specifies the reason for failure (access denied or lack of capacity).

Both the device and the coordinator can issue a disassociation request to end the association.

When a device loses its association with its parent (e.g., it has been moved out of range), it sends orphan notifications (a frame composed of a MAC header, followed by the orphan command code). If it accepts the reassociation, the coordinator should send a realignment frame that contains the PAN ID, coordinator short address, and the device short address. This frame can also be used by the coordinator to indicate a change of PAN ID.

1.3.3   802.15.4 Addresses

1.3.3.1   EUI-64

Each 802.15.4 node is required to have a unique 64-bit address, called the *extended unique identifier* (EUI-64). In order to ensure global uniqueness, device manufacturers should acquire a 24-bit prefix, the *organizationally unique identifier* (OUI), and for each device, concatenate a unique 40-bit *extension identifier* to form the complete EUI-64.

In the OUI, one bit (M) is reserved to indicate the nature of the EUI-64 address (unicast or multicast), and another bit (L) is reserved to indicate whether the address was assigned locally, or is a universal address (using the OUI/extension scheme described above).

1.3.3.2   16-Bit Short Addresses

Since longer addresses increase the packet size, therefore require more transmission time and more energy, devices can also request a 16-bit short address from the PAN controller.

The special 16-bit address FFFF is used as the MAC broadcast address. The MAC layer of all devices will transmit packets addressed to FFFF to the upper layers.

1.3.4   802.15.4 Frame Format

The MAC layer has its own frame format, which is described in Figure 1.9.

The type of data contained in the payload field is determined from the first 3 bits of the frame control field:

- *Data frames* contain network layer data directly in the payload part of the MAC frame.
- The *Ack frame* format is specific: it contains only a sequence number and frame check sequence, and omits the address and data fields. At the physical layer, Ack frames are transmitted immediately, without waiting for the normal CSMA/CA clear channel assessment and random delays. This is possible because all other CSMA/CA transmissions begin after a minimal delay, leaving room for any potential Ack.
- The payload for *command frames* begins with a command identifier (Figure 1.10), followed by a command specific payload.

In its desire to reduce frame sizes to a minimum, 802.15.4 did not include an upper-layer protocol indicator field (such as Ethertype in Ethernet). This now causes problems, since both ZigBee and 6LoWPAN can be such upper layers.

1.3.5   Security

802.15.4 is designed to facilitate the use of symmetric key cryptography in order to provide data confidentiality, data authenticity and replay protection. It is possible to use a specific key for each pair of devices (link key), or a common key for a group of devices. However, the mechanisms used to synchronize and exchange keys are not defined in the standard, and left to the applications.

The degree of frame protection can be adjusted on a frame per frame basis. In addition, secure frames can be routed by devices that do not support security.

1.3.5.1   CCM* Transformations

802.15.4 uses a set of security transformations known as CCM* (extension of CCM defined in ANSI X9.63.2001), which takes as input a string "a" to be authenticated using a hash code and a string "m" to be encrypted, and delivers an output ciphertext comprising both the encrypted form of "m" and the CBC message authentication code (CBC MAC) of "a". Figure 1.11 shows the transformations employed by CCM*, which uses the AES block cipher algorithm E.

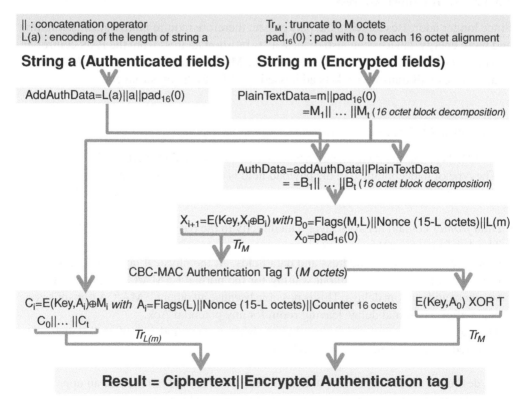

Figure 1.11   Overview of CCM* security transformations.

Security control field	Security attributes	Data confidentiality (data in "m" string)	Data authenticity (data in "a" string)
'000'	None	OFF	No
'001'	MIC-32	OFF	MHR, Auxiliary security header, Nonpayload fields, Unsecured payload fields
'010'	MIC-64	OFF	
'011'	MIC-128	OFF	
'100'	Encrypted fields	Unsecured payload fields	No
'101'	Encr. Fields+MIC-32		MHR, Auxiliary security header, Nonpayload fields
'110'	Encr. Fields +MIC-64		
'111'	Encr. Fields +MIC-128		

Figure 1.12   Security control field codes.

In the case of 802.15.4, $L = 2$ octets, and the nonce is a 13-octet field composed of the 8-octet address of the device originating the frame, the 4-octet frame counter, and the one-octet security-level code.

1.3.5.2   The Auxiliary Control Header

The required security parameters are contained in the *auxiliary control header*, which is composed of a security control field (1 octet), a frame counter (4 octets) ensuring protection against replay attacks, and a key identifier field (0/1/5 or 9 octets).

The first 3 bits of the security control field indicate the security mode for this data frame (Figure 1.12), the security mode determines the size of M in the CCM* algorithm (0, 4, 8 or 16 octets), and the data fields included in the "a" and "m" strings used for the computation of the final ciphertext (security attributes). The next 2 bits indicate the key identifier mode (Figure 1.13) and the remaining bits are reserved.

Key identifier mode	Description	Key Identifier field length
'00'	Key determined implicitly from the originator and recipient of the frame	0
'01'	Key is determined from the 1-octet Key-index subfield of the Key identifier field, using the MAC layer default Key source	1
'10'	Key is determined explicitly from the 4-octet Key source subfield, and the 1-octet Key index subfield of the Key identifier field (part of the auxiliary security header)	5
'11'	Key is determined explicitly from the 8-octet Key source subfield, and the 1-octet Key index subfield of the Key identifier field (part of the auxiliary security header)	9

Figure 1.13   Key identifier mode codes.

1.3.5.3 Key Selection

802.15.4 does not handle distribution of keys: the interface between the MAC layer and the key storage is a *key lookup* function, which provides a lookup string parameter that is used as an index to retrieve the appropriate key.

The lookup material provided depends on the context (see Figure 1.13):

* With implicit key identification (KeyIdMode = "00"), the lookup data is based on the 802.15.4 addresses. The design implies that, in general, the sender indexes its keys according to destinations, and the receiver indexes its keys according to sources.

Addressing mode	Sender lookup data (based on *destination* addressing mode)	Receiver lookup data (based on *source* addressing mode)
Implicit	Source PAN short or extended address	Destination PAN short or extended address
Short	Destination PAN and destination node address	Source PAN and destination node address
Long	Destination node 802.15.4 8 octet extended address	Source node 802.15.4 8 octet extended address

* With explicit key identification, the lookup data is composed of a key source identifier, and a key index. The design implies that the key storage is organized in several groups called key sources (one of which is the *macDefaultKeySource*). Each key source comprises several keys identified by an index.

The CCM standard specifies that a given key cannot be employed to encrypt more than 2^{61} blocks, therefore the applications using 802.15.4 should not only assign keys, but also change them periodically.

1.4   Uses of 802.15.4

802.15.4 provides all the MAC and PHY level mechanisms required by higher-level protocols to exchange packets securely, and form a network. It is, however, a very constrained protocol

* It does not provide a fragmentation and reassembly mechanism. As the maximum packet size is 127 bytes (MAC layer frame, see Figure 1.7), and the MAC headers and FCS will take between 6 and 19 octets (Figure 1.9), applications will need to be careful when sending unsecured packets larger than 108 bytes. Most applications will require

security: the security headers add between 7 and 15 bytes of overhead, and the message authentication code between 0 and 16 octets. In the worst case, 77 bytes only are left to the application.

- Bandwidth is also very limited, and much less than the PHY level bitrate of 250 kbit/s. Packets cannot be sent continuously: the PHY layer needs to wait for Acks, and the CSMA/CA has many timers. After taking into account the PHY layer overheads (preamble, framing: about 5%) and MAC layer overheads (between 15 and 40%), applications have only access to a theoretical maximum of about 50 kbit/s, and only when no other devices compete for network access.

With these limitations in mind, 802.15.4 is clearly targeted at sensor and automation applications. Both ZigBee and 6LoWPAN introduce segmentation mechanisms that overcome the issue of small and hard to predict application payload sizes at the MAC layer. An application like ZigBee takes the approach of optimizing the entire protocol stack, up to the application layer for use over such a constrained network. 6LoWPAN optimizes only the IPv6 layer and the routing protocols, expecting developers to make a reasonable use of bandwidth.

1.5   The Future of 802.15.4: 802.15.4e and 802.15.4g

In the last few years, there has been an increased focus on the use of 802.15.4 for mission critical applications, such as smart utility networks (SUN). As a result, several new requirements emerged:

- The need for more modulation options, notably in the sub-GHz space, which is the preferred band for utilities who need long-range radios and good wireless building penetration.
- The need for additional MAC layer options enabling channel hopping, sampled listening and in general integrate recent technologies improving power consumption, resilience to interference, and reliability.

1.5.1   802.15.4e

Given typical sensor networks performance and memory buffers, it is generally considered that in a 1000-node network:

- Preamble sampling low-power receive technology allows one message per node every 100 s;
- Synchronized receive technology allows one message per node every 33 s;
- Scheduled receive technology allows one message per node every 10 s.

Working group 15.4e was formed in 2008 to define a MAC amendment to 802.15.4:2006, which only supported the last mode, and on a stable carrier frequency. The focus of 802.15.4e was initially on the introduction of time-synchronized channel hopping, but in time the scope expanded to incorporate several new technologies in the 802.15.4 MAC layer. 802.15.4e also corrects issues with the 802.15.4:2006 ACK frame (no addressing information, no security, no payload) and defines a new ACK frame similar to a normal data frame except that it has an "ACK" type. The currently defined data payload includes time-correction information for synchronization purposes[1] and optional received quality feedback.

Some of the major new features of 802.15.4e are described below.

1.5.1.1 Coordinated Sampled Listening (CSL)

Sampled listening creates an illusion of "always on" for battery-powered nodes while keeping the idle consumption very low. This technology is commonly used by other technologies, for example, KNX-rf. The idea is that the receiver is switched on periodically (every macCSLperiod, for about 5 ms) but with a very low duty cycle. On the transmission side, this requires senders to use preambles longer than the receiving periodicity of the target, in order to be certain that it will receive the preamble and keep the receiver on for the rest of the packet transmission. For a duty cycle of 0.05% and assuming a 5-ms receive period, the receive periodicity (macCSLperiod) will be 1 s, implying a receive latency of up to 1 s per hop. CSL is the mode of choice if the receive latency needs to be in the order of one second or less.

In 802.15.4e, CSL communication can be used between synchronized nodes (in which case the preamble is much shorter and simply compensates clock drifts), or between unsynchronized nodes in which case a long preamble is used (macCSLMaxPeriod). The latter case occurs mainly for the first communication between nodes and broadcast traffic: the 802.15.4e ACK contains information about the next scheduled receive time of the target node, so the sender can synchronize with the receiver and avoid the long preamble for the next data packet, as illustrated in Figure 1.14.

802.15.4e CSL uses a series of microframes ("chirp packets", a new frame type introduced in 15.4e) as preamble. The microframes are composed of back-to-back 15.4 packets, and include a rendezvous time (RZtime) and optional channel for the actual data transmission: receivers need to decode only one chirp packet to decide whether the coming data frame is to their intention, and if so can decide to go back to sleep until RZtime and wake up again only to receive the data frame.

CSL supports streaming traffic: a frame-pending bit in the 15.4e header instructs the receiver to continue listening for additional packets.

---

[1] The value, in units of approximately approximately 0.954 μs, reports the PDU reception time measured as an offset from the scheduled start time of the current timeslot in the acknowledger's time base.

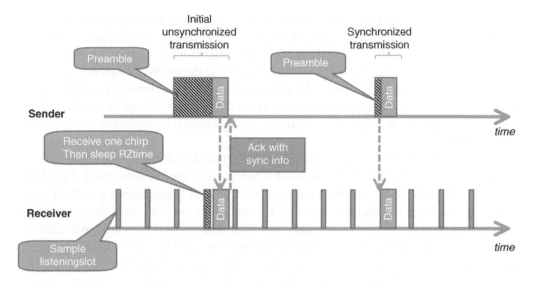

Figure 1.14   Overview of 802.15.4e CSL mode.

1.5.1.2   Receiver-Initiated Transmission (RIT)

The RIT strategy is a simple power-saving strategy that is employed by many existing wireless technologies: the application layer of the receiving node periodically polls a server in the network for pending data. When using the RIT mode, every macRitPeriod, the receiver broadcasts a datarequest frame and listens for a short amount of time (macRit-DataWaitPeriod). The receiver can also be turned on for a brief period after sending data.

The downside of this approach is that the perceived receive latency is higher than in the CSL strategy, and multicast is not supported (must be emulated by multiunicast). The polling typically takes about 10 ms, so in order to achieve an idle duty cycle of 0.05% the macRITPeriod must be 20 s. RIT is adapted to sensor applications, which can tolerate long receive latency.

1.5.1.3   Time-Synchronized Channel Hopping (TSCH)

Channel hopping is a much-awaited feature of 802.15.4:

- It adds frequency diversity to other diversity methods (coding, modulation, retransmission, mesh routing), and will improve the resilience of 802.15.4 networks to transient spectrum pollution.
- In a multimode network, there are situations in which finding a common usable channel across all nodes is challenging. With channel hopping, each node to node link may use a specific set of frequencies.

Channel hopping is supported in the new ACK frame, which contains synchronization information. In an uncoordinated peer to peer network, the channel hopping penalty is only for the initial transmission, as the sender will need to continue to send "chirp packets" on a given send frequency until it becomes aligned with the receiver frequency. After the first ACK has been received, the sender and the receiver are synchronized and the sender will select the sending frequency according to the channel schedule of the receiver. If all joined nodes are in sync, then synchronizing to a single node is enough to be synchronized to the whole network.

The time-synchronized channel hopping (TSCH) mode defined by 802.15.4e defines the operation model of a 802.15.4e network where all nodes are synchronized. The MAC layer of 802.15.4e nodes can be configured with several "slotframes", a collection of timeslots repeating in time characterized by the number of time slots in the cyclical pattern, the physical layer channel page supported, and a 27-bit channelMap indicating which frequency channels in the channel page are to be used for channel hopping. Each slotframe can be used to configure multiple "links", each being characterized by the address list of neighboring devices connected to the link (or 0xfff indicating the link is broadcasting to everyone), a slotframeId, the timeslot within the slot frame that will be used by this link, the channel offset of the link,[2] the direction (receive, transmit or shared), and whether this link should be reported in advertisement frames. Each network device may participate in one or more slotframes simultaneously, and individual time slots are always aligned across all slotframes.

The FFD nodes in a TSCH mode 802.15.4 network will periodically send advertisement frames that provide the following information: the PAN ID, the channel page supported by the physical layer, the channel map, the frequency-hopping sequence ID (predefined in the standard), the timeslot template ID[3] (predefined in the standard), slotframe and link information, and the absolute slot number[4] of the timeslot being used for transmission of this advertisement frame. The advertisement frames are broadcast over all links configured to transmit this type of frame.

For PANs supporting beacons, synchronization is performed by receiving and decoding the beacon frames. For nonbeacon-enabled networks, the first nodes joining the network synchronize to the PAN coordinator using advertisement frame synchronization data, then additional nodes may synchronize to existing nodes in the network by processing advertisement frames. For networks using the time division multiple access mode, where precise synchronization of the whole network is essential, a new flag "clockSource" in the FFD state supports the selection of clock sources by 802.15.4e nodes without loops. A keep-alive mechanism is introduced to maintain synchronization.

---

[2] Logical channel selection in a link is made by taking (absolute slot number + channel offset) % number of channels. The logical channel is then mapped to a physical channel using predefined conventions.

[3] The timeslot template defines timing parameters within each timeslot, e.g TsTxOffset=2120 μs, TsMaxPacket=4256 μs, TsRxAckDelay=800 μs, TsAckWait=400 μs, TsMaxAck=2400 μs.

[4] The total number of timeslots that has elapsed since the start of the network.

1.5.2 802.15.4g

IEEE task group 802.15.4g focuses on the PHY requirements for smart utility networks (SUN).

802.15.4g defines 3 PHY modulation options:

- Multiregional frequency shift keying (MR-FSK): providing typically transmission capacity up to 50 kbps. "Multiregional" means that the standard maps a given channel page to a specific FSK modulation (2GFSK, 4GFSK ...), frequency and bitrate. The current draft contains multiple variants for each region, implying that generic 802.15.4g radios will have to be extremely flexible.
- Multiregional orthogonal quadrature phase shift keying (O-QPSK): providing typically transmission capacity up to 200 kbps.
- Multiregional orthogonal frequency division multiplexing (OFDM): providing typically transmission capacity up to 500 kbps.

The number of frequency bands also increases to cover most regional markets:
- 2400–2483.5 MHz (Worldwide): all PHYs;
- 902–928 MHz (United States): all PHYs;
- 863–870 MHz (Europe): all PHYs;
- 950–956 MHz (Japan): all PHYs;
- 779–787 MHz (China): O-QPSK and OFDM;
- 1427–1518 MHz (United States, Canada): MR-FSK;
- 450–470 MHz, 896–901 MHz, 901–902 MHz, 928–960 MHz (United States): MR-FSK;
- 400–430 MHz (Japan);
- 470–510 MHz (China): all PHYs;
- 922 MHz (Korea): MR-OFDM.

802.15.4g is particularly interesting in Europe, where 802.15.4:2006 allowed a single channel (868.3 MHz). 802.15.4g now offers multiple channels:
- from 863.125 to 869.725 MHz in steps of 200 kHz (MR-FSK 200 kHz);
- from 863.225 to 869.625 in steps of 400 kHz (MR-FSK 400 kHz);
- from 868.3 to 869.225 MHz in steps of 400 kHz(O-QPSK);
- from 863.225 to 869.625 MHz in steps of 400 kHz (OFDM).

As the number of potential IEEE wireless standards and modulation options increases, the frequency scanning time would become prohibitively long if a coordinator was to scan all possible channels using all possible modulations. To solve this problem and improve coexistence across IEEE standards, 802.15.4g defines a new coex-beacon format, using a standard modulation method that must be supported by all coordinators (the common signaling mode or CSM defined in 802.15.4g).

1.5.7 802.15.4g

IEEE task group 802.15.4g focuses on the PHY requirements for smart utility networks (SUN).

802.15.4g defines 3 PHY modulation options.

2

Powerline Communication for M2M Applications

Paul Bertrand
Technology Consultant

2.1 Overview of PLC Technologies

For decades, powerline communication technologies (PLC) have made it possible to use power lines to send and receive data. This "no-new-wire" approach makes PLC one of the best communication technology candidates for the Smart Grid, compared to other wired technologies. On the other hand, as PLC technologies use a media that was not specified for communication, they have faced a number of technical challenges limiting diffusion to niche indoor markets or dedicated ultralow rate applications.

More recently, the booming of modern modulation techniques in integrated silicon made it possible to improve both communication reliability and data rate. Combined with the versatility of emerging protocols such as 6LoWPAN (see the 6LoWPAN chapter), a much larger market is opening for PLC.

Instead of offering here a detailed description of the modulation techniques in use by different vendors/alliances, this can be found for example in [1], this section is more focused on the evolution and comparison of emerging technologies, in the context of the specific requirements of M2M communication.

2.2 PLC Landscape

This section presents an overview of existing powerline technologies and standards. It is not exhaustive and focuses on the most widespread technologies.

The Internet of Things: Key Applications and Protocols, First Edition.
Olivier Hersent, David Boswarthick and Omar Elloumi.
© 2012 John Wiley & Sons, Ltd. Published 2012 by John Wiley & Sons, Ltd.

2.2.1   The Historical Period (1950–2000)

This first period was driven mainly by utilities for outdoor applications at very low frequencies and with an extremely low rate.

The first experiment started in 1950 for remote street lighting. Basically it was one-way On/Off signaling of 10 kW switches at 10 Hz.

In the **mid-1980s** research began on the use of electrical distribution grids to support data transmission, in the [5–500 kHz] frequency band, always for one-way communication.

In **1989**, the ST7536 was the first monolithic half-duplex synchronous FSK modem suitable for applications according to EN 65 065-1 CENELEC and FCC specifications.

In **1997** the first tests for bidirectional data signal transmission over the electrical distribution network were conducted. A specific research effort was started by Ascom (Switzerland) and Norweb (UK).

2.2.2   After Year 2000: The Maturity of PLC

In the year 2000 the tremendous development of personal computer and home networking triggered more and more demand for high bitrate transmission technologies. As FSK modulation in the CENELEC band suffers strong data-rate limitations, the communication industry, inspired by the boom of ADSL, decided to study the implementation of OFDM (orthogonal frequency-division multiplexing) in the band above 2 MHz.

2.2.2.1   High-Rate Modulations

Homeplug and Homegrid are industrial alliances aiming to publish specifications or white papers on powerline technologies. Since 2000 Homeplug has issued different products standards allowing high data rate communication on existing home electric wires. Usually, these alliances are participating to standard organizations like IEEE and ITU.

Homeplug standards, as all high rate modulation technologies in powerline, use OFDM modulation in the frequency band above 2 MHz.

The typical performance of Homeplug implementations is outlined below:

- HP V1.0: peak PHY rate up to 14 Mb/s.
- HP AV: peak PHY rate up to 200 Mb/s.
- HP AV2 compliant with IEEE 1901: peak PHY rate up to 400–600 Mb/s.
- HP GreenPhy is a low-power profile of IEEE 1901 dedicated to smart grid applications and has a peak PHY rate of 10 Mb/s.

Such impressive performance levels, over such a harsh medium as residential powerline, are only possible through usage of complex signal processing, high power level of PLC

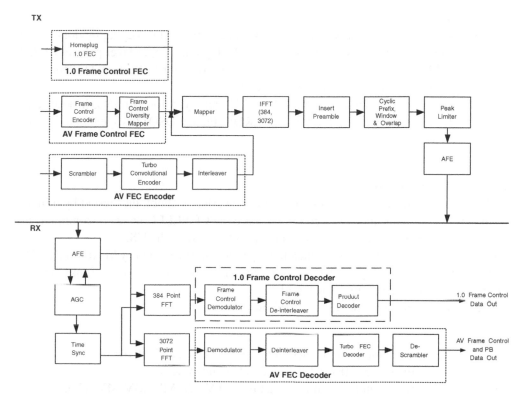

Figure 2.1 HPAV OFDM Transceiver Copyright © 2005, HomePlug® Powerline Alliance, Inc.

emissions and accordingly high power consumption. And still the effective real user data rate usually falls to less than $1/4$ of the PHY maximum data rate in real-life networks.

Figure 2.1, extracted from the HPAV-White-Paper from HomePlug alliance, shows the complexity involved in this type of modem.

2.2.2.2 Low-Rate Modulations

Compared to high-rate OFDM modulation, the complexity of low-rate modems is much lower and a standard DSP can, in most cases, implement the digital processing part of the modem. There are many existing technologies in this field, sometime with a very large installed base.

ISO/IEC 14 908-3 (LonWorks)

This international standard describes an FSK modulation in CENELEC A Band and is used for example in millions of meters in Italy. Refer to Section 4.2 for more details, including the application layer of LonWorks.

IEC 61 334 (S-FSK)

IEC 61 334 is an international standard for low-speed spread shift keying modulation on powerline. It is also known as EN 61 334-5-1:2001. Optional upper layer parts of 61 334 form the DLMS protocol used in metering applications (refer to Section 11.3 for more details on DLMS, including the application layer).

IEC 61 334 is the technology behind G1-PLC, deployed for example, in France as the last-mile communication technology for ERDF (France DSO) first-generation Linky meter.

G3-PLC and PRIME Alliance

G3-PLC and PRIME are two noninteroperable powerline technologies tested by utilities for metering communication. In Europe, both use the CENELEC bands to communicate but they intend also to operate in FCC band up to 500 kHz in the US.

G3-PLC is the powerline communications specification promoted by Maxim and ERDF (French DSO implementing G3 in the second-generation deployment of its "Linky" meter) for smart grid implementations worldwide.

The main specifications of G3 are as follows:

- 10 to 490 kHz operation complies with FCC, CENELEC, and ARIB;
- Coexists with IEC 61 334, IEEE 1901, and ITU G.hn systems;
- Transmission on low- and medium-voltage lines (LV–LV, MV–MV, MV–LV);
- OFDM modulation;
- IEEE 802.15.4-based MAC layer;
- 6LoWPAN adaptation layer supports IPv6 packets;
- AES-128 cryptographic engine;
- Adaptive tone mapping and channel estimation.

In the CENELEC A Band, used by utilities, the OFDM modulation is based on the division of the band into 70 tones. These tones can be modulated in either DBPSK (1 bit per tone), or DQPSK (2 bits per tone).

PRIME is an acronym (PRIME = powerline intelligent metering evolution) for an industry alliance focused on the development of a new open, public and nonproprietary powerline telecom solution for smart grid.

PRIME, created in May 2009, counts more than twenty members at the end of 2010. Principal members are ST Micro, Texas Instruments, ITRON, Landis & Gyr, Iberdrola, Current, ADD and ZIV.

PRIME is based on OFDM multiplexing in CENELEC-A band and is said to reach up to 100 kbps raw data rate. The OFDM signal itself uses 97 (96 data plus one pilot) equally spaced subcarriers with a short cyclic prefix.

Differential modulation schemes are used, together with three possible constellations: DBPSK, DQPSK or D8PSK. Thus, theoretical encoded speeds of around 47 kbps, 94 kbps and 141 kbps (if the cyclic prefix was not considered) could be obtained.

Both G3-PLC and PRIME specifications are the basis of current discussions within IEEE P1901.2 and ITU G.hnem.

Other Low-Rate Modulations

Many other modulations are used for low-rate applications. Most of them use simple modulations like FSK or OOK (X10 for example). Others use different models like spread spectrum or pulse modulation [3].

Table 2.1 shows a comparison between main low-rate technologies in term of standardization, frequency band and modulation.

2.3   Powerline Communication: A Constrained Media

Considering the advantages of powerline technology (no additional wires are required, the network already exists in every home), it is clear that this technology should already be deployed in every home for all types of home-automation applications. However, this is not the case.

We will see that reasons of this limited success are due to four key factors.

2.3.1   Powerline is a Difficult Channel

The channel mandated by CENELEC or FCC for communication is one of the noisiest existing. In homes most appliances now switch in the zero-crossing area of the voltage signal, creating strong spikes. Figures 2.2a and b show some examples of this noise.

Other categories of noise exist: a precise description can be found in [1].

In addition to strong noise levels, the CENELEC band is also known as a strong fading media.

Figure 2.3 shows an illustration of typical fading profiles in the CENELEC A, B, C, D bands.

The notches shown in Figure 2.2 are not stable and could change apparently randomly with time and location.

2.3.2   Regulation Limitations

By definition PLC injects high frequencies in the electric network wires. This injection may induce radio emissions in the HF spectrum and is likely to interfere with existing radio services. For this reason PLC emission and radiation have been regulated from the very beginning.

Table 2.1 Comparison of different powerline technologies

Organization Type	Organization(s)	Technology Name	Frequency Bands	Characterization	Modulation Methods	Signal Level Plan
International SDO	ISO/IEC, ANSI, LonMark	LonWorks, ISO/IEC 14908-3, ANSI 709.2	A (86 kHz and 75.453 kHz 125–140 kHz Fc = 131.579 kHz CENELEC A, B, C	Dual carrier A (86 kHz and 75.453 kHz) C (131.579 kHz and 86.232 kHz)	BPSK/NRZ	EN 50065-1, FCC Part 15
International SDO	ISO/IEC, BS	KNX, ISO/IEC 14543-3-5, EN 50090	PL110 (95–125 kHz Fc = 110 kHz) PL132 (125–140 kHz Fc = 132.5 kHz)	A(60, 66, 72, 76, 82.05 86 kHz) B 110 kHz, C 132.5 kHz	S-FSK/NRZ	EN 50065-1
International SDO	IEC TC57 WG09	IEC 61334-3-1, IEC 61334-5-1, IEC 61334-5-2, IEC 61334-4-32	CENELEC A 20–95 kHz	10 kHz tone separations	S-FSK	EN 50065-1
Industry Specification PRIME Alliance	uSyscom, ADD, STM, TI	PRIME	CENELEC A (~42 –~89 kHz), Capable up to 500 kHz	97 subcarriers, 488 Hz spacing	OFDM/DBPSK, DQPSK, D8PSK	EN 50065-1, FCC Part 15
Industry Specification Public specification	Maxim, ERDF, TI	LF NB OFDM G3 PLC	35.9–90.6 kHz/ CENELEC-A	36 carriers/ CENELEC-A	OFDM/DBPSK, DQPSK	EN 50065-1, FCC Part 15
Industry Specification	INSTEON Alliance	INSTEON	131.65 kHz	Single carrier	BPSK	FCC Part 15

Industry Specification	HomePlug Alliance	HomePlug C&C	FCC (120–400 kHz), CENELEC A,B	Spread Spectrum FCC 120–400 kHz CA 20–80 kHz, CB 95–125 kHz	Single carrier, spread spectrum DCSK6, DCSK4	EN 50065-1, FCC Part 15
Proprietary Specification	PCS	UPB	4–40 kHz		PPM	FCC Part 15
Proprietary Specification	Pico Electronics	X10	120 kHz	Single carrier	OOK	FCC Part 15
Proprietary Specification	ACT	A10	120 kHz	Single carrier	OOK	FCC Part 15
Proprietary Specification	Phillips	Phillips TDA5051A	Fc within 95–145 kHz (132.5 kHz typical)	Single carrier	ASK	FCC Part 15
Proprietary Specification	Ariane Controls	PLM-1	50–500 kHz, 262 kHz expected	Sirgle carrier	FSK/NRZ	FCC Part 15
Proprietary Specification	ENEL	SITRED	CENELEC A 20 kHz–95 kHz	10 kHz tone separations	S-FSK	EN 50065-1
Proprietary Specification	Maxim	G3 Lite MAX2990	10–490 kHz CENELEC-FCC	Adjustable number of Subcarriers	OFDM	EN 50065-1
Proprietary Specification	Watteco	WPC	1.7–4 MHz	LRWBS for low-rate wide band services	Pulse modulation	LRWBS

(Continued)

Table 2.1 (*Continued*)

Organization Type	Organization(s)	Technology Name	Frequency Bands	Characterization	Modulation Methods	Signal Level Plan
Proprietary Specification	Yitran/Rensas	C&C Turbo	FCC (120–400 kHz) CENELEC A,B	Spread Spectrum FCC 120–400 kHz CA 20–80 kHz, CB 95–125 kHz	Single Carrier, spread spectrum, DCSK Turbo, DCSK6, DCSK4	EN 50065-1, FCC Part 15
Technologies under development (no detailed public information available)						
International SDO	IEEE P1901.2	LF NB OFDM (Draft)	CENELEC A/B/C/D FCC 9 to 434 kHz		OFDM	EN 50065-1, FCC Part 15
International SDO	ITU-T SG-15/T4	G.hnem (Draft)	CENELEC A/B/C/D FCC 9 to 434 kHz		OFDM	EN 50065-1, FCC Part 15
International SDO	ISO 15118 CEI TC69 ETSI M2M	Electrical vehicle communication protocol vehicle and Grid	CENELEC A/B/C/D			EN 50065-1
International SDO	ETSI PLT	Low-rate home automation	1.7–4 MHz	LRWBS	OFDM	prEN 50561-1

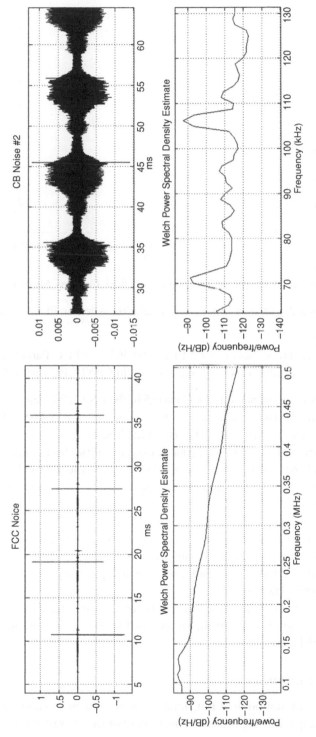

Figure 2.2 (a) and (b): Spikes and noise around the zero-crossing zone. Courtesy Ytran.

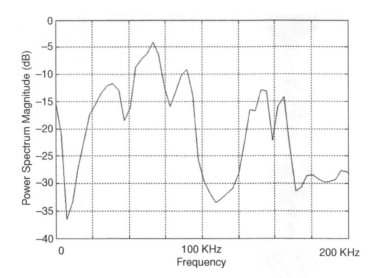

Figure 2.3   Example of absorption in the field. Courtesy Maxim.

Basically there is a distinction in term of regulation whether PLC is narrowband or broadband.

- Narrowband services are subject to CENELEC 50 065-1 or FCC Part15 regulations.
- Broadband services regulation is still in discussion in CENELEC (prEN 50 561-1).

For narrowband services, compared to FCC Part15 in the USA, PLC regulation in Europe appears to be more restrictive as only the frequency band less than 148.5 kHz is available for transmission while it is open up to 500 kHz in the US.

EN 50 065-1 allocates the 3–95 kHz frequency band to utilities for metering applications (it is known as the A band) and reserves the 95–125 kHz, 125–140 kHz and 140–148.5 kHz bands (known in early versions of the standard respectively as B, C, D bands) for analog and digital application within homes, commercial or industrial premises and for control of equipment connected to the low-voltage distribution network. Typical examples of B, C, D band applications are street lighting, electric vehicles or home automation.

As EN 50 065-1 conveys no rights to any user to communicate over any part of the electricity network owned by another party, services using narrowband PLC in home are limited to the 95–148.5 kHz band.

This 15-kHz-limited bandwidth severely limits services to extremely low rates applications and is one of the reasons for the limited diffusion of home automation PLC in homes in the CENELEC bands.

2.3.3   Power Consumption

Until recently, power consumption of PLC modems was not seen as a constraint due to the natural access to energy. But the increased awareness of the overall power consumption of IT-related technologies, has now installed power consumption as a major constraint for new PLC technologies:

- Low power is now mandatory for many technologies including powerline systems for smart grid and smart metering deployments.
- In Europe, the "less than $^1/_2$ Watt" European Directive 2005/32/EC on standby power imposes new paradigms for powerline technologies.

If we look at the numbers: the average consumption of the best IEEE 1901 200 Mb/s powerline modem is around 6 W, while it is close to 4 W for a OFDM CENELEC modem. In a home environment, 4 or 6 W are insignificant compared to a 2 kW air-conditioning unit. But, in the context of M2M with a large number of connected devices, or compared to the consumption of a meter or replacement of a switch it is certainly prohibitive.

It is also prohibitive because the power supply size and heat dissipation are particularly challenging in the context of the form factor required by a meter or a switch.

If we look now at the main contributing factors to the overall PLC technology power consumption, it appears it is roughly balanced between analog subsystems and digital subsystems.

According to Moore's law, digital parts have seen astounding progress in size and performance. But, at the end of the day, having to support increasing data rate, tough channels, sampling frequencies and signal processing are engaged in a never-ending race, thus limiting Moore's Law benefits regarding power consumption.

On the other hand, except for some ultralow-power coupling technologies [3], power consumption of couplers and analog front end (AFE) are limited in their progress by the emission level required by the injection of high-frequency signals in a difficult media.

As a consequence of restrictive regulations and slow progress in power-consumption reduction, some companies are now working on efficient management of sleeping modes and standby states. Radio technologies, like ZigBee or 6LoWPAN, have implemented the same strategy in the past for battery-powered nodes. At first sight, it appears to be an efficient solution for AC powered nodes too . . . but it isn't: sleeping modes unfortunately do not solve the power-supply size problem. It is a common [marketing] mistake to mix up overall peak power consumption and energy. A low-energy system, expressed in watt-hour, could need a relatively high peak power, in watt, when operating in "periodic wake-up" mode, and AC power supplies need to be dimensioned according to peak power requirements.

2.3.4   Lossy Network

As shown in Figure 2.4, the PLC link may be subject to as many disturbances as a wireless link, because every electrical device may inject noise and/or absorb the signal. Considering the number of electrical devices in the electrical network of a typical multidwelling

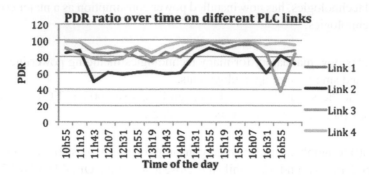

Figure 2.4 Packet delivery ratio (PDR) variation over time on several PLC links. Courtesy Watteco.

unit and their varying electrical behavior that randomly disturbs the communication channel, the routing mechanism over PLC networks has to cope with very lossy links, as well as dynamic loss characteristics. Furthermore, these noise/fading generators create asymmetric links that add routing complexity.

Implementation of special purpose routing protocols is now considered as the solution to reach full coverage with a lossy network and difficult channels. The Internet Engineering Task Force (IETF) recognized the need to form a new Working Group to standardize an IPv6-based routing solution for IP smart object networks, which led to the formation of a new Working Group called ROLL "Routing Over Low power and Lossy" networks in 2008.

Here is the charter of ROLL:

Low-power and lossy networks (LLNs) are made up of many embedded devices with limited power, memory, and processing resources. They are interconnected by a variety of links, such as (IEEE 802.15.4, Bluetooth, low-power WiFi, wired or other low power PLC (powerline communication) links. LLNs are transitioning to an end-to-end IP-based solution to avoid the problem of noninteroperable networks interconnected by protocol translation gateways and proxies.

The routing protocol RPL [2] developed in ROLL is the protocol chosen by ZigBee™ IP and 6LoWPAN. The advantage of RPL is that it is independent of the media and then it can be the basis of interoperability between PLC and radio sensors. Interoperability is simply achieved by routing messages from a PLC node to a wireless node. For more details on RPL, refer to the Section 12.4.

Figure 2.5 shows an illustration of this dual PHY sensor network.

The conclusion is that the powerline medium appears to be a very challenging medium, limited by regulation and power consumption and until recently no existing standard really offered a good alternative to wireless for M2M application in home. Fortunately,

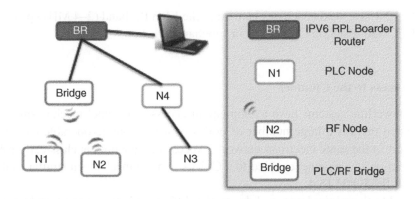

Figure 2.5   Wireless nodes and PLC nodes exchange messages through a RPL border router.

the recent development of ROLL and the requirement of using IPv6 open the door to new powerline standards interoperable with wireless services using the same routing protocol.

2.3.5   Powerline is a Shared Media and Coexistence is not an Optional Feature

2.3.5.1   Why is Coexistence So Important?

This question of coexistence in PLC has been discussed in many PLC standard organizations. Many articles can be found in the Smart Grid Interoperability Panel (http:// collaborate.nist.gov/twiki-sggrid/bin/view/SmartGrid/SGIP).

Powerline cables are a shared medium. They cannot provide links dedicated to a single user and there is no practical way to insulate two neighbors. The signals transmitted within an electrical network can interfere with signals generated within an adjacent house or apartment. It is then likely that these interferences will produce data rate and quality of service drops. The same issues exist in other share media like radio.

For this reason, it is necessary to define mechanisms to limit the harmful interference caused by noninteroperable neighboring devices.

Different mechanisms exist or are proposed today:

- EN 50 065-1 provides a CSMA/CA algorithm and a 132.5-kHz "channel busy" signal for the C band.
- IEEE 1901 and ITU G9972 provide a standard called intersystem protocol (ISP) implementing a TDMA and frequency-division mechanism for a fair access to the shared resource.
- Current discussions in IEEE 1901.2 are also investigating the coexistence between OFDM CENELEC band transmission for metering applications and legacy narrowband services (based on FSK modulation for example).

- prEN50412-4 is a coexistence mechanism in the LRWBS band (2–4 MHz) proposed to CENELEC and based on CSMA/CA in subchannels)

2.3.5.2 Access to the Channel

Low-rate powerline systems have to implement access to channel mechanisms and especially when there is a large number of nodes or when noninteroperable technologies may transmit in the same electrical network. Access to a channel is classically ruled by energy detection in a channel coupled to a CSMA generic mechanism similar to the one described in IEEE 802.15.4.

Compared to the original one, and due to the difference between radio and powerline, some timing constants are different, in particular the backoff period. Basically, the backoff period is the base time a node has to wait before it can transmit after sensing the channel. Compared to 802.15.4 wireless systems, a low-rate PLC modem offers higher latency and lower data rates, however the backoff period has to be chosen short enough (1 ms for example) to allow a sufficient responsiveness for the latency requirements of home-automation applications like switching.

In order to evaluate the impact of data rate and the number of nodes versus average access time, simulations have been carried out with different situations:

- Number of sensors from 5 to 30. All sensors are independent and try randomly to access the network.
- The number of tries/minute is presented vertically.
- A maximum of fail/success rate of 3 is supposed.
- There are no propagation effects and the channel is perfect (best case).

The parameters are:

- backoff period = 1 ms;
- frame size varies randomly from 16 bytes to 128 bytes;
- a failed transmission occurs when the total backoff time > 100 ms (typical minimum latency for home automation HMI).

Figure 2.6 shows that the network supports 20 nodes communicating 400 ms every minute, provided the data rate is a least 100 kb/s.

With a lower data rate, on average every node will retry more than 3 times to access the channel.

Meters transmitting through a powerline modem in the CENELEC band might be impacted by the results of these simulations. For example in a multidwelling situation it is likely that more than 30 nodes might be within reach of the meters. If a high activity level in homes is required (real-time display of the consumption, thermal regulation ...)

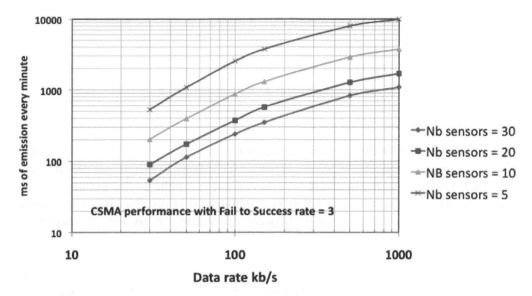

Figure 2.6 Simulations of the impact of number of nodes on access time. Courtesy Watteco.

the 30 nodes could saturate the channel and the latency of the communication between the meter and substations could increase.

Prioritizing meter communication and increasing data rate can minimize this effect but still not cancel it.

Another solution is to use different frequency bands in the home and for metering. For example, the CENELEC band for metering and the LRWBS band (2–4 MHz) for home automation.

2.4 The Ideal PLC System for M2M

PLC combined with recent IPv6 developments could lead to a new standard for wired communication in homes. Recently, in the context of European Mandate 441 for metering applications and home automation, ETSI PLT has defined the requirements of this new standard.

What are the main requirements of this new PLC M2M standard?

- Openness and availability;
- Range;
- Energy consumption;
- Data rate;
- Robustness;
- EMC regulatory compliance;

- Coexistence with other PLC technologies;
- Security;
- Latency;
- Interoperability with M2M wireless services.

2.4.1   Openness and Availability

Open standards, compared to proprietary developments, are accelerating factors for the dissemination and the success of any mass-market technology. Standard openness encourages interoperability, stimulates multisourcing and low-cost policies. The use of available standards, when adaptation to in-home PLC automation is possible, is a tremendous factor of stability and interoperability between existing mature technologies and new emerging ones.

- Home automation standards must use existing, available and internationally recognized PHY, MAC and DLL technologies. Examples are: IEEE 802.15.4, 6LoWPAN, IPv6, RPL . . .

2.4.2   Range

Range (or coverage) is one of the most important requirements in PLC. It is very challenging because of the extreme harshness of the medium. multidwelling units (MDU) where multiphases networks can exceed an internode distance of 100 m are probably the worst-case situation in terms of range performance.

- The standard should be able to cover all outlets of the home. Routing protocols like RPL can be used to ensure full coverage if needed.

2.4.3   Power Consumption

Power consumption is a very sensitive parameter for home automation applications.

- The standard must enable products to comply with local low power regulations like Directive 2005/32/EC on energy consumption in standby mode in Europe or Energy Star in the US.
- The standard must enable products with low power consumption in relation to the savings users may request for energy-efficiency products.
- The power consumption of a powerline node must be comparable to the power consumption of a wireless node.

2.4.4 Data Rate

Data rate is not a critical requirement per se for home automation applications. A data rate of 10 kb/s is, in most cases, enough to cover in-home lighting or switching applications. However, factors like routing protocols, security, access to media mechanisms in a home with dozen of nodes will probably increase communication stream and payload.

For that reason, it is critical not to limit the standard data rate to 10 kb/s. On the other hand, high-rate PLC systems (>1 Mb/s) are not a good compromise for evident reasons of power consumption. 100 kb/s appears to be a good compromise.

* The standard must provide a nominal rate of 100 kb/s in field installations.
* In the case of very noisy channels, the standard should support variable data rates to keep reliable links.

2.4.5 Robustness

The home environment can be very challenging for PLC nodes. Harsh channels in homes are due to various physical reasons:

* Low impedance appliances (from 1 ohm to 10s of ohms, pure capacitive loads, etc.);
* Disturbance from common electric devices (chargers, dimmers, switching power supplies, etc. . . . .);
* Absorption from breakers and ground fault circuit interrupter;
* Electrical topology (multiphase wiring, neutral/ground connections, etc.).

In order to offer a good end-user experience:

* The standard must provide close to 100% connectivity in the home.
* The standard may use routing or/and data rate adaptation in order to keep connectivity in harsh environments.
* The standard must support multiphase topologies, optionally with phase couplers. Across the world, different home wiring topologies exist using single phase (France, Spain . . .), dual phase (US and Japan) or three phases (Germany and Northern countries). Usually, the PLC signal is injected in one phase and natural crosstalk between phases may not always be sufficient. In that case phase couplers ensure reinjection of signal from one phase to the other phases.

2.4.6 EMC Regulatory Compliance

- The standard must comply with local EMC regulations in force in the frequency band in use.

2.4.7 Coexistence

- The standard must implement existing coexistence standards when using frequency bands where other PLC systems are also transmitting.

Coexistence standards already in use by legacy PLC systems are:

- EN 50 065-1 C band: (channel busy signaling at 132.5 kHz);
- ISP mechanism in the 2–30 MHz band (as described in IEEE 1901 or ITU G.9972).

2.4.8 Security

- The standard must provide services to support encryption and secure data services. However, a compromise should be found between security and cost of implementation. Furthermore, plug and play installation may conflict with strong security requirements.
- The security suite must be open and available.

2.4.9 Latency

- The standard must support low-latency communication in conformance with end-user usual expectation in home automation. Usually, a latency of 100 ms is considered as a maximum for home automation applications.

2.4.10 Interoperability with M2M Wireless Services

- The standard must ensure interoperability with 6LoWPAN and other wireless compatible protocol through gateways embedding both nodes and running RPL.

2.5 Conclusion

PLC technologies, after a rather difficult and slow start, are now on track for mainstream deployments triggered by smart grid and home automation investment programs and mandates all over the world.

It is worth noting that new paradigms and business models will deeply influence the specifications of emerging standards. The capacity of the communication channel to support encrypted IPv6 frames, the large number of nodes, low power consumption and interoperability with radio transmission technologies are clearly the next challenges engineers will have to overcome.

References

[1] PowerLine Communications (2010) *Theory and Applications for Narrowband and Broadband Communication Over Power Lines*, Wiley

[2] RPL: IPv6 Routing Protocol for Low power and Lossy Networks - http://tools.ietf.org/html/draft-ietf-roll-rpl.

[3] IPSO White Paper #6 – "A survey of several low power Link layers for IP Smart Objects" by JP Vasseur, Paul Bertrand, Bernard Aboussouan, Eric Gnoske, Kris Pister, Roland Acra and Allen Huotori.

Part Two

Legacy M2M Protocols for Sensor Networks, Building Automation and Home Automation

3

The BACnet™ Protocol

BACnet stands for Data Communication Protocol for **B**uilding **A**utomation and **C**ontrol **Net**works. Unlike most other protocols that began as private implementations followed by standardization efforts, BACnet was built from the ground up as an independent, royalty-free, open standard control and automation protocol. The standard committee was chaired by university professors until 2004, its goal was to harmonize data types and formats, data exchange primitives, and common application services. Several open source BACnet stacks are available.(e.g., http://bacsharp.sourceforge.net/; http://bacnet.sourceforge.net/.).

The scope of BACnet applications is very large, including HVAC (heating, ventilating, and air conditioning) applications, lighting control, fire control and alarm, security, and interfacing to utility companies.

Together with LonWorks, BACnet is one of the most popular industrial automation and control protocol, adopted in products of many leading vendors (Siemens Building Technologies, Johnson Controls, Inc., Teletrol Systems,@IC, TAC, KMC Controls, American Auto-Matrix, Contemporary Controls Ltd, Reliable Controls).

3.1 Standardization

The BACnet standardization effort began in 1987 during a Standard Project Committee meeting of ASHRAE (American Society of Heating, Refrigerating and Air-Conditioning Engineers). BACnet became an ISO standard in 2003 (ISO 16 484-5). In January 2006 the BACnet Manufacturers Association and the BACnet Interest Group of North America combined their operation in a new organization called BACnet International (http://www.bacnetassociation.org/), which provides conformance testing

The Internet of Things: Key Applications and Protocols, First Edition.
Olivier Hersent, David Boswarthick and Omar Elloumi.

services (BACnet Testing Laboratories) and promotes the adoption and development of the standard.

3.1.1   United States

BACnet became a standard in 1995 as ASHRAE/ANSI standard 135 and a conformance testing method was standardized in 2003 as BSR/ASHRAE Standard 135.1. The last revision of the standard was published in 2010.

3.1.2   Europe

BACnet was adopted in 2003 by CEN (Comité Européen de Normalization, http://www.cen.eu) Technical Committee 247, for the management level and automation level. For the Automation level, it coexists with EIBnet (Konnex), at the Field level, CEN adopted Konnex (merger of three European protocols EIB (European Installation bus), Batibus, EHS), and LonWorks/LONTalk.

Europe has a specific European user and interest group: http://www.big-eu.org.

3.1.3   Interworking

BACnet ability to interwork with other technologies has always been a key concern, and BACnet does provide enough flexibility to allow mapping of other common protocols to a BACnet model. However, there are often many ways of providing such a mapping, and there is a need to formally specify a standard mapping in order to ensure interoperability of interprotocol gateway implementations:

- BACnet interoperability with Konnex (KNX) control protocol has been specified in Annex H/5.
- BACnet interoperability with ZigBee has been specified in Annex X.

3.2   Technology

BACnet focuses on the network layer and above. At the presentation layer, it uses ASN.1 syntax[1] for the definition of all data structures and messages (application protocol data units or APDUs). The BACnet transport layer adds routing information to these APDUs,

---

[1] ASN.1, or "Abstract Syntax Notation 1" is defined in ISO/IEC 8824. This syntax, widely used in the telecom world, is used to define precisely data structures, and also functional primitives. It includes a standard serialization mechanism : ASN.1 BER (simple but less efficient), and ASN.1 PER (complex but extremely efficient).

and the resulting messages may be carried on top of virtually any link layer, using the adaptation functions provided by the BACnet network layer.

3.2.1   Physical Layer

BACnet upper layers are independent from the underlying physical layer, facilitating the implementation of BACnet on most popular networks. BACnet physical layers have been defined for ARCNET, Ethernet, IP tunneling (defined for routers interconnecting BACnet segments over IP in Annex H), BACnet/IP (devices are IP aware and can communicate directly over IP networks), RS-232 (BACnet Point to Point), RS-485 (with BACnet specific Master-Slave/Token Passing LAN technology, up to 32 nodes on 1200 m, at 76 kbit/s on shielded twisted pairs), and LonWorks/LonTalk.

Since 2008, there is also a standard implementation of BACnet over ZigBee® (Annex X).

3.2.2   Link Layer

BACnet can be implemented directly on top of the LonTalk or IEEE802.2 (Ethernet and ArcNet) data link layers. It also defines a data link layer (Point to Point PTP) for RS232 serial connections, and a MS/TP data link layer for RS-485.

For IP or other network technologies that can be used as link layers, the standard defines a BACnet virtual link layer (BVLL) that formalizes all the services that a BACnet device might require from the link layer, such as broadcasts.

For instance, BACnet devices may implement the IP BVLL, which encapsulates the required control information not readily available from the native IP link layer (e.g., a flag indicating whether the message was received as a unicast or broadcast), in a BACnet virtual link control information (BVLCI) header (see Figure 3.1). Thanks to the IP BVLL, BACnet devices become full-fledged BACnet IP devices, able to communicate directly over IP without a need for an "Annex H" router. Similarly, a BACnet device could implement an ATM, frame relay or ISDN BVLL in order to become a native node in these networks.

On many link layers, broadcasts are difficult or have their own limitations. BACnet has a concept of a "BACnet broadcast management device" (BBMD), which implements the broadcast requirements of BACnet for the selected link layer, for example, it may convert a BACnet broadcast into IP-based multiunicast and/or broadcast messages. Devices can register with the BBMD to receive broadcast messages dynamically.

3.2.3   Network Layer

BACnet is primarily defined as a network layer protocol, which defines the network addresses required for the routing of messages. BACnet networks consist of one or more

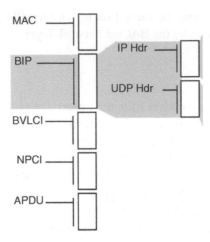

Figure 3.1 Transport of a BACnet application message (APDU) over IP/UDP.

segments consisting of single physical segments or multiple physical segments connected by repeaters. The BACnet segments are connected by bridges if they employ the same LAN technologies, or BACnet routers otherwise.

BACnet addresses are hierarchical: the formal separation of the network identifier and the address identifier simplifies routing. Addresses have a variable length, which makes it easy for BACnet to adapt to the native addresses of underlying link layers (Figure 3.2 shows the BACnet use of an IP address). The BACnet network header (NPCI) can include the following information elements:

– A 2-byte source network (SNet) and variable length source address (SAddr, SLen). For Ethernet, ArcNet and MS/TP, the native protocol address format is used, for LonTalk,

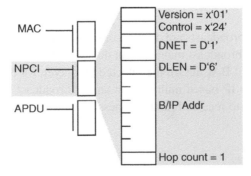

Figure 3.2 BACnet message from a BACnet non-IP device to a BACnet/IP device.

the concatenation of the subnetID and nodeId (2 bytes), or the concatenation of the subnetID and Neuron ID (7 bytes) is used.

- A 2-byte destination network (DNet) and variable-length destination address (Daddr,Dlen). For broadcast messages, DNet identifies the network on which a broadcast is required, and Dlen=0.
- A 4-level network priority indicator
- A 1-byte message type: 9 message types are used by the BACnet routing mechanisms (e.g., *Who-is-router-to-network* to discover a router to a specific networkID). Vendors can define specific extension message types.
- A 2-byte vendor ID.

Not all information elements have to be present, depending on the specific use-case: a *control* bitmask field specifies which information elements are present.

3.2.4 Transport and Session Layers

BACnet implements a collapsed OSI model in which the transport and session layers are not required. The application layer provides the required reliability mechanisms usually associated with the transport layer, as well as the segmentation and sequencing mechanisms usually associated with the session layer.

3.2.5 Presentation and Application Layers

BACnet does not attempt to formally separate the presentation layer and the application layer (a separation that is often a bit artificial for most protocols anyway). BACnet models the various features of devices as *objects*, exchanging *service* primitives. The service primitives are described using ASN.1 syntax and serialized using ASN.1 BER (basic encoding rules, ITU-T Recommendations X.209 and X.690, for a good introduction on ASN.1 and BACnet, see http://bacnetbill.blogspot.com/2009/10/bacnet-tagging-rules.html).

3.2.5.1 BACnet Objects

BACnet abstracts the device basic functions as *objects*: each *device* is decomposed into a collection of standardized objects, where physical inputs and outputs and other characteristics of the object (name, type, configuration parameters) are represented by *properties*. Each object is identified by a unique *Object_Identifier* within the device. See Table 3.1 for a list of standard BACnet objects.

BACnet currently lists 30 object types, for which it defines standard properties and the expected behavior:

```
BACnetObjectType ::= ENUMERATED {
        access-door              (30),
        accumulator              (23),
        analog-input             (0),
        analog-output            (1),
        analog-value             (2),
        averaging                (18),
        binary-input             (3),
        binary-ouput             (4),
        binary-value             (5),
        calendar                 (6),
        command                  (7),
        device                   (8),
        event-enrollment         (9),
        event-log                (25),
        file                     (10),
        group                    (11),
        life-safety-point        (21),
        life-safety-zone         (22),
        load-control             (28),
        loop                     (12),
        multi-state-input        (13),
        multi-state-output       (14),
        multi-state-value        (19),
        notification-class       (15),
        program                  (16),
        pulse-converter          (24),
        schedule                 (17),
        – see averaging          (18),
        – see multi-state-value  (19),
        structured-view          (29),
        trend-log                (20),
        trend-log-multiple       (27),
        – see life-saftey-point  (21),
        – see life-saftey-zone   (22),
        – see accumulator        (23),
        – see pulse-converter    (24),
        – see event-log          (25),
        – enumeration value 26 is reserved for a future addendum
        – see trend-log-multiple (27),
        – see load-control       (28),
        – see structured-view    (29),
        – see access-door        (30),
```

Table 3.1   Standard BACnet objects

AnalogInput
AnalogOutput
AnalogValue
BinaryInput
BinaryOutput
BinaryValue
Calendar
Command
Device
EventEnrolment
File
Group
Loop
MultistateInput
MultistateOutput
NotificationClass
Program
Schedule
Averaging
MultistateValue
TrendLog
LifeSafetyPoint
LifeSafetyZone
Accumulator
PulseConverter

- **Device:** all devices are required to implement the device object. The *Object_Identifier* of the device object must be unique across the BACnet network and is the identifier of the physical device implementing that device object. All other objects are implemented only if needed. The device object lists all objects implemented by the device.
- **Binary input, Binary output, Binary value**.
- **Analog input, Analog output, Analog value, Averaging** (function that monitors a signal and records its min, max and average value).
- **Multistate input, Multistate output, Multistate value**.
- **Accumulator:** for devices implementing a feature that counts pulses. **Pulse converter**: counts pulses or takes an accumulator object as an input, but the count can be offset (adjusted) at any time, and scaled.
- **Loop:** properties modeling a feedback control loop.
- **LifeSafetyPoint** (such as smoke detectors, pull stations, sirens...), **LifeSatefyZone** (properties associated to a group of LifeSafetyPoints and LifeSafetyZones).

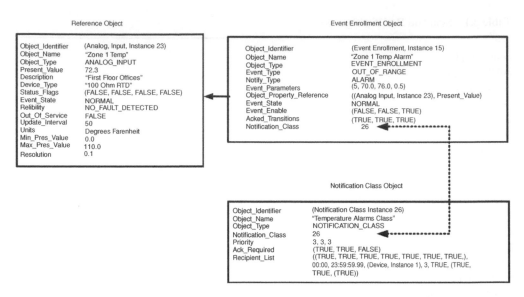

Figure 3.3   Example use of BACnet objects for event notification (from ANSI 135 standard document).

- **Access door:** object modeling a physical door.
- **Calendar** (list of dates). **Schedule**: the schedule object describes a periodic schedule (days and time of day, with exceptions), and for each period associates time-dependent values to the properties of other objects.
- **Event enrollment:** this object defines the required conditions for an event to occur (e.g., a change of state or value, a command failure, a value getting off range for a certain duration, a sudden change of a value, etc.), and which devices or objects must be notified if this happens. See Figure 3.3.
- **Notification Class:** this list of objects that are enrolled for an event notification is used as a property of the Event enrollment object. It also contains priority parameters. The list of notified elements may vary according to days of the week and time of day. See Figure 3.3.
- **Command** (a command object can be used to execute a set of actions changing properties of other objects).
- **File:** lists the properties of a file object that may be accessed using File services.
- **Program:** list the properties modeling an application program.
- **Trend Log:** this object models a function that logs, at periodic intervals, a given property value. Such a log can be triggered based on predefined conditions. **Trend Log Multiple Objects**: same as a Trend Log, but stores multiple property values in parallel. **Event Log**: "FIFO" log of timestamped event notifications.
- **Group:** a group of objects. **Structured View:** organizes other objects in an org-chart-type structure.

- **Load control:** this object provides the interfaces for power-load shedding, for example, processing of requests from a utility company. It support hierarchical load shedding, scheduled shedding, compliance reporting.

Typical object properties include the object name, a reliability indicator (no error detected, no sensor, etc.), the present value of a counter and its unit, a scaling factor, maximum and minimum values. They also include configuration parameters for object features, for example, how long a value must remain out of range before reporting an error condition, or the minimal increment of a real value triggering an update notification.

3.2.5.2   BACnet Services

BACnet considers that all objects are servers that provide *services*. It defines 5 classes of services, the description of each service can be found in ANSI/ASHRAE 135 clauses 13 to 17 (see Table 3.2).

Each service uses a set of messages supporting the related communication needs. The messages are defined using ASN.1 syntax (ANSI/ASHRAE 135 clause 21) and exchanged using standard remote operation primitives (request, indication, response, confirm):

- *Alarm and event services:* BACnet provides multiple event reporting options: objects may support "intrinsic reporting" (e.g., report an event periodically, report error conditions, status updates), or may be configured by means of *Event enrollment* object (Figure 3.3) to report specific conditions such as a change of value (*COV reporting*), or a value out of range. The latter mechanism, called *algorithmic reporting* implements a subscribe-notify model for events. The objects that requested to be notified are listed in *Notification Class* objects (Figure 3.3).

  The following service primitives are defined for event management and reporting: **AcknowledgeAlarm** (self-explanatory), **ConfirmedCOVNotification** (*Change of value* event notification primitive in which receivers must report the success or failure of actions taken as a result of the event), **UnconfirmedCOVNotification**, **ConfirmedEventNotification**, **UnconfirmedEventNotification**, **GetAlarmSummary** (BACnet events can be flagged as alarms, in which case a list of active alarms is returned by this primitive), **GetEnrollmentSummary** (returns a list of event-notifying objects according to specified filters, such as objects with an active event enrollment from another object), **GetEventInformation** (returns a list of active event states within a device), **LifeSafetyOperation** (e.g., silence a siren), **subscribeCOV** (subscribe to *Change of value* notifications for an object), **SubscribeCOVProperty** (subscribe to *Change of value* notifications for a property).
- *File access services:* read and write primitives are atomic, that is, a single operation is executed at a time.

Table 3.2   Standard BACnet services

Who Is
I Am
Who Has
I Have
Read Property
Write Property
Device Communication Control
ReinitializeDevice
Atomic Read File
Atomic Write File
Time Synchronization
UTC Time Synchronization
Subscribe COV
Subscribe COV Property
Confirmed COV Notification
Unconfirmed COV Notification
Read Property Multiple
Read Property Conditional
Read Range
Write Property Multiple
Get Alarm Summary
Get Event Information
Get Enrollment Summary
Acknowledge Alarm
Confirmed Event Notification
Unconfirmed Event Notification
Unconfirmed Text Message
Confirmed Text Message
Add List Element
Remove List Element
Create Object
Delete Object
Unconfirmed Private Transfer
Confirmed Private Transfer
VT Open
VT Data
VT Close
Life Safety Operation
Get Event Information

- *Object access services:* a set of self-explanatory primitives: **ReadProperty, ReadPropertyConditional, ReadPropertyMultiple, WriteProperty, WritePropertyMultiple, CreateObject, DeleteObject, AddListElement, RemoveListElement**.
- *Remote device management services:* a set of primitives for maintenance purposes (start and stop BACnet message transmission, send vendor specific commands, reinitialize a device, time synchronization). Among those primitives, the *Who-Has/I-Have* services are used to discover which devices on the network have a given object name or object ID, the *Who-Is/I-Am* primitives are used to discover devices on a BACnet network.
- *Virtual terminal services:* primitives for bidirectional exchange of character-oriented data, "Telnet like".

3.3   BACnet Security

BACnet device A supporting security can request a session key from a key server for a future communication with device B. The key server will generate a session key SKab and transmit it securely to A and B (encrypted with the private keys of A, respectively B). BACnet uses 56-bit DES encryption.

Device A may then authenticate a future transaction with B: A and B authenticate each other by exchanging challenges (based on random numbers encrypted with the session key), the challenge message includes the identifier (InvokeID) of the future transaction to be authenticated.

A may also ensure the confidentiality of the future transaction by encrypting the corresponding application message with the session key.

3.4   BACnet Over Web Services (Annex N, Annex H6)

The XML working group of ASHRAE SSPC 135 has introduced Addendum c to BACnet-2004 that specifies a Web Services interface to building automation and control systems. The addendum is in two parts:

- Annex N to BACnet defines the BACnet Web Services interface, BACnet/WS. BACnet/WS is a connectionless protocol using a Simple Object Access Protocol 1.1 interface (http://www.w3.org/2002/ws/) over HTTP (RFC 2616). The model and primitives exposed in Annex M are independent from the underlying protocol and could apply to any building automation protocol (LonWorks, KNX, ModBus . . .).
- Annex H6, Combining BACnet Networks with Non-BACnet Networks, that prescribes the gateway mapping specifically to and from BACnet messages. Annex H6 exposes a specific profile for Web Services access to underlying BACnet objects.

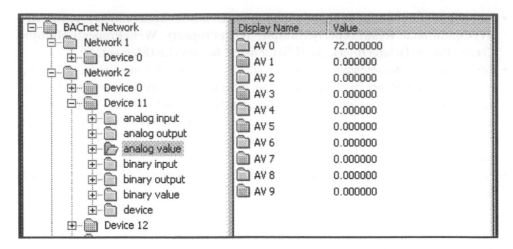

Figure 3.4 Snapshot of the SCADA engine BACnet web service.

3.4.1   The Generic WS Model

The BACnet/WS fundamental data structure is a tree of *nodes* under a single *root node*. In order to allow more flexibility in the tree structure, BACnet/WS has a notion of *reference node* that can serve as an alias to a *referent node*.

Each node and each property is identified by a path:

– /Floor2/Room3/Discharge Temp identifies a node;
– /Floor2/Room3/Discharge Temp:inAlarm identifies property inAlarm of node DischargeTemp;
– Another format commonly used by BACnet WS vendors (Figure 3.4) is the following /[Network]/[Device]/[ObjectType]/[Instance], for instance **/2/11/2/0** is a path to **network 2, device 11, analog value (object type 2) instance 0**;
– Special value <Path>:Children indicates all the nodes one level below the current node. For instance /:Children returns all the children of the root path. /1/1/0:Children returns all paths to Analog Inputs on Device 1.

The network visible state of each node is exposed as the node value, and a collection of attributes (Figure 3.5). BACnet uses Primitive attributes (native XML types), Array attributes (arrays of Primitive values), Enumerated attributes (choice of XML strings specified by BACnet/WS). Only the Value attribute is writable with the services defined by the standard.

BACnet/WS defines a small set of 5 standard nodes required in any implementation, for example:10.5

.sysinfo/.software-version, a string containing the software revision of the software running on the server

Attribute	XML type	Description
NodeType	String	Hint about a node content. One of Unknown /System /Network /Device /Functional /Organizational /Area /Equipment /Point /Collection /Property /other
NodeSubType	String	
Description	String	
DisplayName	String	
Aliases	String	
Reference		
Attributes	String	Array containing the list of all attributes present in this node.
Children	String	Array containing the collection of identifiers for the children of this node when accessed through this path. The path to the each child is obtained by concatenation of the specified child identifier with the path of the current object.
HasDynamic Children	Boolean	
Value	Depends on ValueType	e.g. if the ValueType is 'OctetString', the XML type of the Value property will be base64binary.
ValueType	String	One of None /String /OctetString /Real /Integer /Multistate /Boolean /Date /Time /DateTime /Duration
Units	String	If ValueType is 'Real' or 'Integer', Engineering unit for the Value attribute of the node, expressed as the enumeration identifier of the corresponding unit in the BACnetEngineeringUnits ASN.1 production (if the canonical service option is used, otherwise arbitrary units can be used)
ValueAge	Double	In seconds
HasHistory	Boolean	
Writable	Boolean	True if the value is writable through web services.
WritableValues	String	Array containing all possible values that may be written to the Value attribute (when ValueType is Multistate or Boolean)
InAlarm	Boolean	True to indicate an alarm condition
PossibleValues	String	Array containing all possible values of the Value attribute (when ValueType is Multistate or Boolean)
MinimumLength	nonNegativeInteger	
MaximumLength	nonNegativeInteger	
Resolution	Depends on ValueType	
Maximum	Depends on ValueType	
Minimum	Depends on ValueType	
Overridden	Boolean	

Figure 3.5   BACnet/WS mandatory and most common attributes.

3.4.2   BACnet/WS Services

BACnet/WS defines the services used to access and manipulate the data on the server:

– **getValue/getRelativeValues:** from a path parameter, the getValue service retrieves a single value for a single attribute on a single node (Figure 3.6). The value can be a primitive type or an array attribute (in which case the results are concatenated with a semicolon separation in the result string). **getValues**: accepts multiple paths parameters and returns multiple values. **getArray/getArrayRange**: these services accept a single path to an array parameter and return an array of strings (the entire array or a portion for getArrayRange. **getArraySize** returns the size of an array.

Request
```
<?xml version="1.0" encoding="UTF-8"?>
<SOAP-ENV: Envelope
 xmlns:SOAP-ENV="http://schemas.xmlsoap.org/soap/envelope/"
 xmlns:SOAP-ENC="http://schemas.xmlsoap.org/soap/encoding/"
 xmlns:xsi="http://www.w3.org/2001/XMLSchema-instance"
 xm"lns:xsd="http://www.w3.org/2001/XMLSchema"
 xmlns:ns="urn:bacnet_ws">
 <SOAP-ENV:Body SOAP-ENV:encodingstyle=
      "http://schemas.xmlsoap.org/soap/encoding/">
  <ns:getvalue>
   <ns:optionsx/ns:options>
   <ns:path>/.sysinfo/.vendor-name</ns:path>
  </ns:getva"lue>
 </SOAP-ENV:Body>
</SOAP-ENV:Envelope>
```

Response
```
<?xml version="1.0" encoding="UTF-8"?>
<SOAP-ENV:Envelope
 xmlns:SOAP-ENV="http://schemas.xmlsoap.org/soap/envelope/"
 xmlns:SOAP-ENC="http://schemas.xmlsoap.org/soap/encoding/"
 xmlns:xsi="http://www.w3.org/2001/XMLSchema-instance"
 xmlns:xsd="http://www.w3.org/2001/XMLSchema"
 xmlns:ns="urn:bacnet_ws">
 <SOAP-ENV:Body SOAP-ENV:encodingstyle=
      "http://schemas.xmlsoap.org/soap/encoding/">
  <ns:getvalueResponse>
   <ns:result>SCADA Engine</ns:result>
  </ns:getvalueResponse>
 </SOAP-ENV:Body>
</SOAP-ENV:Envelope>
```

Figure 3.6   Example WS getValue request and response for the value of .sysinfo/.vendor-name.

- **setValue:** sets a new value for a single attribute on a single node identified by a path. **setValues** sets multiple values identified by multiple paths.
- **getHistoryPeriodic:** specifies a sampling interval and a start time for a property identified by its path, and returns a specified number of samples interpolated according to a specified method.
- **getDefaultLocale/getsupportedLocales:** retrieves the locale(s) that the server has configured for its default locale/supports.

Some services accept service options that modify their behavior or their return values: readback (reads back the value is the result string of setValue), errorString, errorPrefix, locale, writeSingleLocale, canonical, precision, noEmptyArrays, writePriority.

3.4.3   The Web Services Profile for BACnet Objects

This profile specifies the mapping of some BACnet object properties to BACnet/WS node attributes (Figure 3.7).

BACnet object property	WS node attribute
Object_Name	DisplayName
Present_Value	Value
Units	Units
Status_flags[IN_alarm]	InAlarm
Action_Text	PossibleValues/WritableValues
Active_Text, Inactive_Text	PossibleValues/WritableValues

Figure 3.7   Some recommended mappings of BACnet object properties to BACnet/WS node attributes.

3.4.4   Future Improvements

3.4.4.1   Updated BacNet/WS Annex N: An ATOM Interface

As this book was going to press, Addendum 2010 a.m. to BACnet was about to be released. It contains a major update of the BACnet/WS interface specification. The major features are:

- The adoption of the ATOM Publishing protocol (defined by IETF RFC 5023) for the REST version of the interface.
- The adoption of the PubSubHubbub subscription model for data push services. PubSubHubbub was defined as a simple subscribe/notify extension to ATOM and RSS. See http://code.google.com/p/pubsubhubbub/ for more details.
- A comprehensive XML representation of BACnet structures (Annex Q), the control system modeling language (CSML).

With this new version, BACnet/WS ceases to be a simplified interface serving limited purposes: it really becomes a fully functional interface that provides access to the full functionality of BACnet.

3.4.4.2 Profile Names

Work is ongoing within the "Applications" (AP) working group, to investigate the development of "macro" object types suitable for various application areas. This group is studying a proposal for a "BACnet modeling language" that is expected to provide a machine-readable way of representing the capabilities of individual BACnet devices such as services and objects supported. A new standard object property called "Profile_Name" now allows the extension of standard (or proprietary) object types in such a way that devices with knowledge of the named profile are able to interpret the extended properties. This is expected to form the basis for convenient BACnet interfaces to other protocols that support object-oriented representation of their functionality, for example, LonWorks.

4

The LonWorks® Control Networking Platform

The LonWorks series of networking protocols were developed by Echelon® Corporation for the needs of control and automation applications and are now managed by the LonMark® International trade group. The LonWorks platform was initially developed in an effort to move away from the proprietary centralized control model, where a central controller receives all measurements from remote sensors and sends all commands to remote actuators. In an effort to eliminate the controller as a single failure point and increase the efficiency and power of control systems, the LonWorks platform introduced a concept of "connection" enabling devices to exchange data directly, using a subscribe/notify model.

At the physical layer, the LonWorks platform is media independent; including media types for copper pairs (wires) and power lines, radio, infrared light, and optical fiber.

The LonWorks platform is one of the most popular protocols for building and industrial automation, claiming over 90 million installed devices.

4.1 Standardization

In 2008, LonWorks also was approved as ISO standards: ISO/IEC 14 908-1, -2, -3, and -4 for the protocol, twisted-pair channel, power-line channel, and IP-tunneled channel, respectively.

4.1.1 United States of America

The communication protocol (a.k.a., the LonTalk® protocol; Echelon's trade name) was submitted to ANSI in 1999 and accepted as a standard for control networking

The Internet of Things: Key Applications and Protocols, First Edition.
Olivier Hersent, David Boswarthick and Omar Elloumi.
© 2012 John Wiley & Sons, Ltd. Published 2012 by John Wiley & Sons, Ltd.

(ANSI/CEA-709.1; originally EIA-709.1). Shortly after, the power line and twisted-pair physical layers were accepted as part of the ANSI standard series.

4.1.2 Europe

The European Committee for Standardization (CEN) standardized the protocol for "buildings" use in 2005. Then in 2007, the LonWorks platform became part of the Application Interworking Specification (AIS) recognized by the European Committee of Domestic Equipment Manufacturers (http://www.ceced.eu/) for Household Appliances Control and Monitoring.

4.1.3 China

In China the LonWorks platform is both a national standard in the category of "controls" (GB/Z 20 177.1-2006) and in the category of "buildings" (GB/T 20 299.4-2006).

4.2 Technology

Figure 4.1 shows the structure of a LonWorks packet, including data fields used by each protocol layer. The protocol layers are detailed in the following sections.

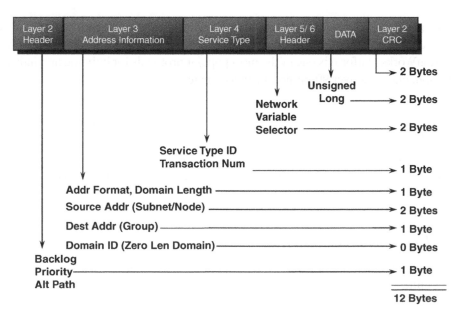

Figure 4.1 A typical ISO/IEC 14 908-1 Packet. Reproduced by permission of © Echelon Corporation.

4.2.1 Physical Layer

The LonWorks protocol is media independent, and assumes only a physical layer that can transmit binary signals, called a channel. Specific transceivers are required for each underlying physical layer. The series of standards defines transceivers for twisted pair, link power, power line, radio frequency, optical fiber, coaxial cable, and infrared media channels (Figure 4.2, the complete list can be found at http://www.lonmark.org/spid). Most of the transceiver channels use differential Manchester encoding, where each "1"

Channel Name	Media	Bit rate and capacity	Reference
FO-20L/ FO-20S	Fiber optic	1.25 Mbps	ANSI/CEA-709.4
IP-852	**LonWorks over IP**	**Over 10000 packets per second**	**ANSI/CEA-852**
PL-20 (L-N)/ PL-20 (L-E)	**126 kHz or 140 kHz BPSK, line to neutral or line to earth**	**5 kbps**	
PL-20A	**CENELEC A-band power line (75 kHz or 86 kHz BPSK)**	**2613 bps (about 11 packets per second)**	**ANSI/CEA-709.2**
PL-20C/ PL-20N	**CENELEC C-band Power line (115 or 132 kHz BPSK) with/without access protocol**	**156.3 k/3987 bps (about 18 (20C) to 20 (20N) packets per second)**	**ANSI/CEA-709.2**
TP/FT-10	**Free topology twisted pair (can be branched at any point) Max wire length : about 500m (900 to 2700m with bus termination). Devices per link : 64 (when power provided by twisted pair), 128 (if devices have a separate device power source)**	**78.13 kbps [(about 180 packets per second, peaks up to 225 packets per second)**	**ANSI/CEA-709.3**
TP/RS485-39	RS-485 Twisted Pair	39.06 kbps	EIA/TIA-232-E
TP/XF-1250	**Transformer-Isolated Twisted Pair**	**1.25 Mbps(about 576 packets per second, peaks up to 720 packets per second)**	**LONMARK® Interoperability Guidelines**

Figure 4.2 LonWorks media channels (most common in bold). Reproduced by permission of © Echelon Corporation.

msb						lsb	
Bitsync (configurable number of 1 bits)						0	
Link Header (1 byte)							
Network Header (4 to 16 bytes)							
Transport Header (0 to 1 bytes)							
Session Header (0 to 1 bytes)							
Presentation Header (1 to 2 bytes)							
Application Data (0 to 246 bytes)							
Link CRC (2 bytes)							

Figure 4.3 A typical LonWorks data frame, least significant bits are transmitted first. Reproduced by permission of © Echelon Corporation.

is transmitted as a polarity reversal for a full period, and each "0" is represented as two polarity reversals during a single, full period. This type of encoding ensures that there is no continuous component in the transmission (it averages to 0 regardless of the information transmitted), and that connections – particularly those using two wires – that not need to care about polarity.

Each physical communication link may be interconnected by means of a LonWorks router, or extended by means of a physical layer repeater. Channels connected by a repeater form a *segment*.

4.2.2   Link Layer

The protocol's link layer provides cyclical redundancy check (CRC) error checking in order to detect most transmission errors; an access, collision avoidance and priority mechanism; and a data-frame format (Figure 4.3).

4.2.2.1   Access and Priority Mechanism

The media access control (MAC) algorithm employed is carrier-sense, multiple access (CSMA), in a variant called p-persistent CSMA: A LonWorks networking device is required to establish that the transmission medium is idle before it can start communicating (this is common to all CSMA protocols). In addition, in order to reduce the probability of collisions, it will begin to transmit, with probability p, in one of $1/p$ predefined time slots (called *beta-2* slots, during typically from 2 to 30 bit times). The number of time slots is dynamically adjusted based on the network load: with more time slots (smaller p), the network works better during high loads, but this adds to the transmission delay compared to fewer time slots. Each LonWorks networking packet includes the number of acknowledgments expected as a result of sending this packet, which allows receiving devices to estimate the upcoming network load and adjust the number of *beta-2* slots accordingly. Adjustments are made as multiples of 16 ($n \times 16$), where n is called the

msb							lsb
Bit-sync (configurable number of 1 bits)							0
Pri	Path		Delta Backlog				
Network Layer Datagram (6 to 246 bytes)							
CCITT CRC-16 (2 bytes)							

Figure 4.4 Details of the link layer header. Reproduced by permission of © Echelon Corporation.

current transmission-channel *backlog*. The required increment is indicated in the link header delta backlog field (Figure 4.4).

Some transceivers can send over two channels for redundancy purposes; the desired channel for a packet is indicated by the *Path* bit of the link header (Figure 4.4).

On each channel, a fixed number of the first *beta-2* time slots (up to 127 time slots) can be allocated to priority packets. Devices can send both priority packets and nonpriority packets. The Pri bit (Figure 4.4) of the link layer header indicates whether the packet is a priority packet.

4.2.3 Network Layer

The network layer provides the message-delivery mechanisms.

Each device ("node") is identified by a unique 48-bit identifier, called the unique node identifier (UID) or the unique_node_ID, within device memory structures. It is also, colloquially and historically, known as the neuron ID, or *neuronID*. The UID does not change over the lifetime of the device. It is normally used only when the device is first inserted in the network, before it has been assigned a logical network address. This facilitates the replacement of a device by a new device of the same type, which will have a different UID but will be assigned the same logical network address as the replaced device. The UID is also utilized for applications requiring authenticated messaging service (for higher-security needs).

The protocol uses hierarchical addressing, and defines the *domain (0, 1, 3, or 6 bytes)*, *subnet (8 bits)*, and *node (7 bits)* subaddresses. Each device is assigned a unique *nodeID* in each subnet. Therefore there may be up to 32 385 devices (255 subnets × 127 nodes) per domain. The devices of a single domain or subnet may be on various channels; and devices from multiple domains may coexist on the same channel.

The source of each message is contained in the address field of the header, and specifies the sending node *subnetID* and *GroupID* (first two bytes). The target of a message may be, depending on the header Addr field value:

- A single node: the header comprises a 2-byte destination address specifying the *subnetID* and *nodeID* (header Addr format=2) or a 7-byte address specifying the destination *subnetID* and *neuronID* (header Addr format field=3).
- All devices in a subnet (3-byte address, header Addr format field=0).
- All devices in a domain (3-byte address, *subnetID*=0, header Addr format field=0).

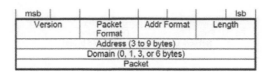

Figure 4.5 The network header format. Reproduced by permission of © Echelon Corporation.

- A group: the protocol defines group addresses (2-byte *domainID*, and 1-byte *groupID*, header Addr format header field=1), so there may be up to 256 groups per domain. Such addresses may be used to address groups of devices on different subnets. There is a maximum of 64 devices per group for acknowledged device-to-device messaging services and no limit for unacknowledged messaging services.

The packet format field specifies whether the packet is a transport packet (packet format field value = 0), a session packet (1), an authenticated packet (2) or a presentation packet (3).

4.2.4   Transport Layer

The protocol's transport layer provides the end-to-end reliability mechanisms. The protocol offers four basic types of messaging service, depending on the desired tradeoff between reliability and efficiency:

- *Acknowledged* (header transport packet format field = 0, see Figure 4.6): messages are sent in the context of a transaction identified by a *transactionID*. Each receiver sends an acknowledgment message (header transport packet format field = 2) with the *transactionID*. If not all acknowledgements have been received (until a configurable timeout), the message is retransmitted with the same transaction ID.
- *Request/response*: the request/response service is managed by the session layer.
- *Repeated* (header transport packet format field = 1): each message is repeated several times so that the probability of a device failing to receive one of the messages is reduced. However, the target devices do not acknowledge these messages, so the service is not fully reliable. Echelon Corporation estimates that 3 repeats results in a successful delivery probability greater than 99.999%. This service is useful for group addressing to large groups.

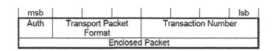

Figure 4.6 Transport layer header details. Reproduced by permission of © Echelon Corporation.

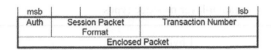

Figure 4.7   The session layer header format. Reproduced by permission of © Echelon Corporation.

- *Unacknowledged*: the message is sent as "best effort" only and the sender is not notified if the message is lost *en route*. This service is useful for periodic data reporting from sensors.

The device can select any of the mechanisms listed above to transport its presentation layer messages.

For authentication to work (Auth bit in the transport header set to 1, see Figure 4.6), a 48-bit key (one per domain) must be configured in each device sending or receiving authenticated messages. The authenticated message is sent as a normal message from A to B but before acting on this message, B will authenticate the message by challenging Λ. The challenge includes a random number and A is expected to reply with a hash (encrypted encoding) of the secret domain key and the challenge. B computes the same hash locally and compares the result with the challenge response of A. If they are identical, B successfully acknowledges the original message.

4.2.5   Session Layer

The session layer replaces the transport layer when the packet format field of the network layer is set to 1 (Figure 4.5). It offers authentication (see transport layer) and a request/response service.

Like the acknowledged messaging service, the request/response service is useful when a message is sent to a device or group of devices and individual responses are required from each receiver. The request message (session packet format=0, see Figure 4.7) may either be resent until all responses (session packet format=2) have been received (the transaction number provides the acknowledgment mechanism), or it may be duplicated several times to minimize the risk of packet loss. The responses from a request/response transaction, unlike the simple, low-level acknowledgment from an acknowledged messaging service, usually include application-level response data.

The enclosed packet data are formatted as a presentation-layer message.

4.2.6   Presentation Layer

4.2.6.1   Presentation-Layer Messages

The presentation layer defines the data-interpretation conventions of the protocol: it uses *messages* that are transported and retransmitted by the lower layers. Except for specific

Table 4.1 LonWorks presentation-layer message types

Message Type	1 byte Message code	Usage
User Application Message	00-2F	Message payload includes a 6-bit message code, followed by data. The applications exchanging application messages must agree on the interpretation of the message codes.
Standard Application Message	30-3E	Same as User application messages but using the standard message codes used for standard application-layer services (data log, file transfer, and self-installation functions).
Foreign-Frame Message	40-4E	Arbitrary data, which may encapsulate other protocols.
Network Diagnostic Message	50-5F	
Network-Management Message	60-7F	
Network Variable Message	80-FF	Identifier that identifies the data as a data value (or data structure) of 1 to 31 bytes that may be shared by multiple devices on a network.

needs, most applications typically exchange data using network variable messages, except for some specific needs (file transfer, self-installation, etc.) or when there are communications requirements beyond those specified by the network variable messages.

The LonWorks networking presentation-layer messages begin with a one-byte message code that defines the type of data contained within the message (Table 4.1), followed by 0 to 277 bytes of data.

4.2.6.2 Network Variables and the LonWorks Subscribe/Notify Model

Network variables are essential interfaces of most LonWorks networking devices. Network variables (Figure 4.8) have a direction (input to receive data, output to send data), type (scalar or aggregate of several fields), length, a self-documentation string. They are identified by a *network variable index* on the device, and are identified by a 14-bit *network variable selector* (0-3FFF) over the network (maintained in a configuration table on each device). A single network variable index can be associated to several network variable selectors on the same device; in which case, one is called the primary network variable

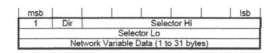

Figure 4.8 Format of a LonWorks network variable message (carried by the transport/session layer). Reproduced by permission of © Echelon Corporation.

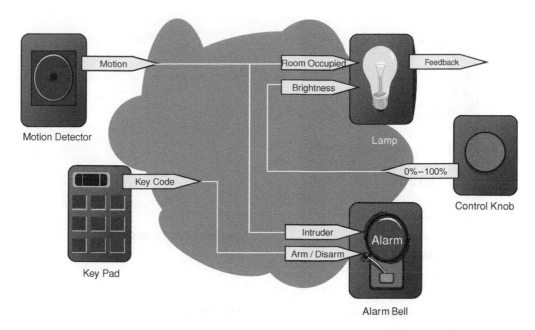

Figure 4.9 Example of devices exchanging information through connections. Reproduced by permission of © Echelon Corporation.

selector and the others are aliases. Network variables' values are exchanged over the network by network variable messages (Figure 4.10).

The LonWorks protocol provides two primitives to poll network variables over a network: one of which can poll multiple devices simultaneously using the network variable selector and one targeted to a single device using the network variable index.

The protocol also has a native subscribe/notify model – a key feature that supports the claim that the networking platform facilitates networks designed without central points of failure. Network variables belonging to different devices can be connected if they have the same type and length, for example, a state output (type: switch-type) of a switch device may be connected to the switch-type input of a lamp. There are several types of connections:

- Unicast: single output to single input;
- Multicast out: single output to multiple inputs on several devices;
- Multicast in: multiple outputs on several devices to a single input.

Connections are created by a process called *binding* and are performed by a network-management tool or by the self-installation process of the device. When binding network variables, the protocol implementation of the device is configured with:

- The list of addresses of the other devices (or groups of devices) in the network expecting that network variable's value;
- The target network variable selectors.

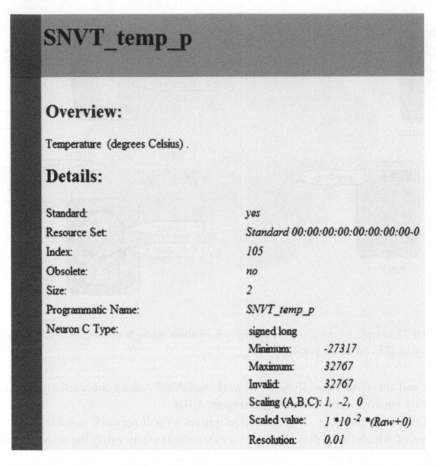

SNVT_temp_p

Overview:

Temperature (degrees Celsius).

Details:

Standard:	*yes*
Resource Set:	*Standard 00:00:00:00:00:00:00:00-0*
Index:	*105*
Obsolete:	*no*
Size:	*2*
Programmatic Name:	*SNVT_temp_p*
Neuron C Type:	signed long
	Minimum: *-27317*
	Maximum: *32767*
	Invalid: *32767*
	Scaling (A,B,C): *1, -2, 0*
	Scaled value: $1 * 10^{-2} * (Raw+0)$
	Resolution: *0.01*

Figure 4.10 Example specification of a LonMark standard network-variable type (SNVT_temp_p). Reproduced by permission of © Echelon Corporation.

The application on the device will only update the network variable, and the protocol implementation will ensure that it is sent to all configured target addresses (domain/ subnet/node or group, using the appropriate network variable selector). See Figure 4.9 for an example network configuration showing several connections using network variables.

The network variable selector values 3000 to 3FFF hex are reserved for unbound network variables, with the selector value equal to 3FFF hex minus the network variable index. Selector values 0 to 2FFF hex are available for bound network variables. This provides a total of 12 288 network variable selectors for bound network variables. Each device can have up to 8192 network-variable aliases and 4096 of those bindable network aliases.

Standard network variable types (SNVTs, pronounced "SNIV-its"), specify standard data encodings (units, range, resolution, scaling, data structure, etc.) covering most common usage cases. The list of SNVTs includes over 200 types and covers a wide range of applications. The complete list is available at types.lonmark.org.

If an application requires a network variable type that is not a SNVT, device manufacturers can define custom network-variable types. These are called *user network variable types (UNVTs)*.

4.2.7 Application Layer

The network configuration and network-diagnostic services are defined by the protocol standard. The following list summarizes the standard application-layer services. Additional standard application-layer services (Figure 4.11) are published at www.lonmark.org

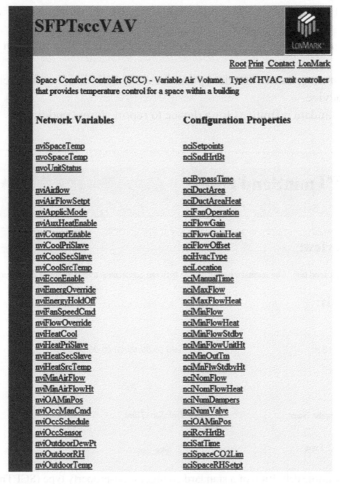

Figure 4.11 Example specification of a LonMark functional profile: the space comfort controller – variable air-volume controller. Reproduced by permission of © Echelon Corporation.

- *Network configuration* – configuration of the network attributes of a device (network address and binding information for the device's network variables).
- *Network diagnostics* – diagnostics commands.
- *File transfer* – the largest practical amount of data that can be transferred in a single packet is 228 bytes, but the LW-FTP file-transfer method transfers data using a stream of 32-byte packets.
- *Application configuration* – provides a standard interface to configure the behavior of a device. The interface is based on configurable data values called *configuration properties*. Standard configuration property types (SCPTs, pronounced "SKIP-its"), are the configuration equivalent to SNVTs. An up-to-date list of SCPTs can be found at types.lonmark.org.
- *Application specification* – documentation of a device's functions as a set of function blocks (a distinct set of complementary network variables and configuration properties).
- *Application diagnostics* – standard testing primitives for function blocks and devices.
- *Application management* – standard primitives to enable, disable, and override function blocks on a device.
- *Alarming* – standard primitives for a device to report alarm conditions.

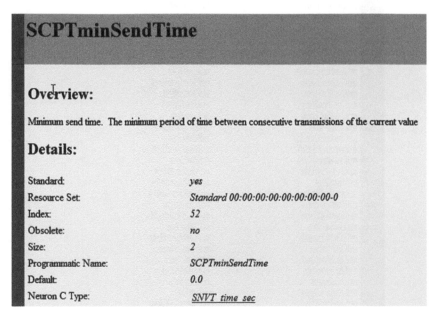

Figure 4.12 Example definition of a standard configuration-property type (SCPTminSendTime). Reproduced by permission of © Echelon Corporation.

- *Scheduling* – standard primitives for scheduling events based on time of day, day of week, and date.
- *Time and date management* – standard primitives for synchronizing the time-of-day and date for devices within a network.

Some application-layer services are defined for device developers through a committee of those developers within LonMark International, the trade association in support of the protocol standards. The culmination of those application-specific functional definitions are known as *functional profiles*, or simply *profiles* (Figure 4.11 and 4.12), and are implemented in part or in their entirety by developers as *function blocks* within the device.

4.3   Web Services Interface for LonWorks Networks: Echelon SmartServer

The Echelon SmartServer is an Internet gateway and local computing platform (e.g., edge control node) that connects LonWorks, ModBus, M-Bus, local I/O to other devices and networks, in addition to providing a SOAP Web Service-based interface for data access and configuration. Besides using SOAP (a W3C recommendation), the SmartServer interface is not formally standardized.

The SmartServer provides access to LonWorks networking devices through *data points*, which contain a value, data type, and format properties. The gateway implements the following functions that can be used to directly access data points: **DataPointRead, DataPointWrite**. For each write, it is possible to specify whether the value must be immediately propagated to the LonWorks network or not. This makes it possible to set several items in a structure sequentially and to propagate all values to the network at once; in an atomic way.

The SmartServer gateway also provides a set of applications accessible though SOAP (Figure 4.13).

The *DataServer* application can be used to create, manage, delete, and access data points. The *Datalogger* can sample and store data-point values in logs and circular buffers. The *AlarmGenerator* generates alarms – for instance, when certain limits of data-point values are exceeded – while the *AlarmNotifier* logs alarm conditions, sends notifications via SMTP e-mail, or updates specific data points. The *EventScheduler* and *EventCalendar* applications can be used for periodic updates of data points. The *TypeTranslator* can be utilized to translate values of data points with a specific variable type into a different type.

4.4   A REST Interface for LonWorks

As we have seen in the previous section, a web services interface already exists for LonWorks. However, the new trend in M2M architecture is to use a REST model, where

DataServer	Datalogger:	AlarmGenerator:	AlarmNotifier:
List	List	List	List
Get	Get	Get	Get
Set	Set	Set	Set
Delete	Read	Delete	Delete
Read	Clear		Read
Write	Delete		Write
ResetPriority			Clear

AnalogFB:	EventScheduler:	EventCalendar:	TypeTranslator:
List	List	List	List
Get	Get	Get	Get
Set	Set	Set	Set
Delete	Delete	Delete	Delete
			RuleList
			RuleGet
			RuleSet
			RuleDelete

Figure 4.13 SmartServer applications and functions. Reproduced by permission of © Echelon Corporation.

the types of interactions are restricted to only the CRUD[1] verbs. This requires a specific design of the representation of the underlying M2M network or device.

Echelon Corporation published a first version of a REST interface for LonWorks in July 2010. In this version 1.0, the "LonBridge Proxy Server REST API" supports the basic CRUD REST interactions, but did not introduce yet a subscribe/notify model: this type of interaction is feasible in a REST architecture but requires both interface sides to act as client and server. Standardized subscribe/notify REST models have been introduced for instance by ETSI TC M2M (refer to Chapter 14 this book): it is expected that ETSI M2M interfaces to LonWorks will also exist in the near future, and introduce the additional interactions and features made possible by the ETSI TC M2M REST architecture.

4.4.1   LonBridge REST Transactions

4.4.2   Requests

The LonBridge API supports the following HTTP request methods:

- GET: retrieve resource data from the LonBridge server;
- POST: create a new resource;

---

[1] Create, read, update, delete.

- PUT: update an existing resource managed by the LonBridge proxy server;
- DELETE: delete a resource.

The specifications for each resource describe how the commands are applied.

4.4.3 Responses

Responses include a response body and a status code.

The response body may be formatted as JSON, XML, HTTP, or text. The default is JSON. The response format may be specified as a suffix to the URL – for example **GET server/api/devices.xml** returns a list of all devices in XML format. The response format may also be specified in the accept header.

The status code is a standard HTTP status code. Typical status codes include the following:

> 200 – OK (standard response for successful request);
>
> 201 – Created (standard response after successfully creating a resource);
>
> 400 – Bad Request (request has invalid syntax or cannot be fulfilled);
>
> 404 – Not Found (requested resource could not be found but may be available in the future);
>
> 500 – Internal Server Error (generic error message when other messages don't apply);
>
> 501 – Not Implemented (request not recognized).

4.4.4 LonBridge REST Resources

The various LonBridge REST resources made available by the API are regrouped in seven functional groups addressing the following domains: network, device, device type, connections, groups and measures.

4.4.4.1 Network

These resources allow to access or modify the main LonWorks network parameters.

Syntax	**http://**_server_[_:port_]**/api/network/**{_resource_} [**?**_params_]
Methods	PUT, GET
Resources	Network resources: **name, domainId, domainLength, key**

4.4.4.2   Devices

Device resources are used to retrieve and update resources on an individual device or a set of devices. The device ID is the LonBridge device ID, which is the letter "o" followed by an identifier, for example: **o0**, **o1**, or **o2**.

Syntax	**http://**_server_[_:port_]**/api/devices**[/{_id_}[/{_resource_[=_value_]}]][?_params_]
Methods	PUT, GET, DELETE
Resources	Device resources: **name**, **type**, **location**, **scene**, **active**, and data points defined per device type: • **blinds: angle, level, motion, scene** • **dimmer** (Lamp Module) resources: **brightness, state (on, off), power, energy, scene** • **switch** (Appliance Module) resources: **state (on, off), power, energy, scene** • **occupancy: occupied** • **thermostat: fan (auto, on), humidity, mode (auto, heat, cool, off), schedule, setback, setpoint, temperature, message, pricing** Device parameters: **startDate** (default current date; specified as _day_[-_month_[-_year_]] where _month_ defaults to current month and _year_ defaults to current year), **startTime** (default current time; 24-h time), **interval** (default 60 minutes), **maxCount** (default 100), and **deviceType** (default all). When a **startDate** or **startTime** parameter is specified, up to **maxCount** records may be returned. Each record includes a timestamp in "_year-month-day hour:minute:second_" format, for example: **2010-07-05 15:43:10**.

Examples

GET server.com/api/devices – returns list of all devices. The following is an example XML encoded response body:

```
<devices>
      <o0 type="switch" name="Appliance Module 1"
           brand="Echelon" active="true" state="on"
           power="27" energy="8.5913" />
      <o1 type="dimmer" name="Lamp Module 1" brand="Echelon"
           active="true" brightness="93" state="on" power="58"
           energy="0.5219" />
</devices>
```

The following is an example JSON encoded response body:

```
{
    "o0": {
        "type": "switch",
        "name": "Appliance Module 1",
        "brand": "Echelon",
        "active": "true",
        "state": "on",
        "power": 27,
        "energy": 8.5913
    },
    "o1": {
            "type": "dimmer",
        "name": "Lamp Module 1",
        "brand": "Echelon",
        "active": "true",
        "brightness": 93,
        "state": "on",
        "power": 58,
        "energy": 0.5219
    }
}
```

GET server.com/api/devices?deviceType=“switch” – returns a list of all switch devices. This corresponds to the LonBridge **<get TBD />** command.

GET server.com/api/devices/o2 – returns a list of all resources defined for device o2. This corresponds to the LonBridge **<o2.get/>** command.

GET server.com/api/devices/o2/power – returns the last power-consumption reading for device o2. This corresponds to the LonBridge **<o2.get select=“state”/>** command.

PUT server.com/api/devices/o2/state – turns on device o2 on or off (the state is sent in the request body). This corresponds to the LonBridge **<o2.set state=“on”/>** command.

DELETE server.com/api/devices/o2 – deletes device o2. This corresponds to the LonBridge **<o2.delete/>** command.

4.4.4.3 Device Types

Lists devices by device type.

Syntax	**http://**_server_[**:**_port_]**/api/**{_deviceType_}**/**{_id_}**/**{_resource_}[**?**_params_]
Methods	PUT, GET, DELETE
Resources	Device resources: same as for **devices.**
	Device parameters: same as device parameters.

4.4.4.4 Connections

Syntax	**http://**_server_[**:**_port_]**/api/connections/**{_id_}**/**{_resource_}[**?**_params_]
Methods	PUT, GET, POST, DELETE
Resources	Connection resources: **state** and **setting**.

4.4.4.5   Scenes

Syntax	**http://***server*[*:port*]**/api/scenes/**{*id*}**/**{*resource*}[**?***params*]
Methods	GET, POST
Resources	Scene resources: **state** and **setting**.

4.4.4.6   Groups

Syntax	**http://***server*[*:port*]**/api/devices/**{*id*}**/groups/**{*id*}**/**{*resource*} [**?***params*]
	http://*server*[*:port*]**/api/groups/**{*id*}**/**{*resource*}[**?***params*]
Methods	PUT, GET, DELETE
Resources	Group resources for devices: **membership** (**true** or **false**).
	Group resources: **state.**

4.4.4.7   Measures

Syntax	**http://***server*[*:port*]**/api/devices/**{*id*}**/groups/**{*id*}**/**{*resource*} [**?***params*]
	http://*server*[*:port*]**/api/groups/**{*id*}**/**{*resource*}[**?***params*]
Methods	GET, POST (define new measure), DELETE (delete measure)
Resources	Measure resources are used to retrieve aggregate calculations for current and
	historical data.

Examples
GET server.com/api/measures – returns list of all measures.
GET server.com/api/measures/energy – returns aggregate energy usage.
GET server.com/api/measures/energy?startDate=1 – returns up to 100 aggregate energy usage historical values since the first of the month.
GET server.com/api/measures/energy?category=lighting – returns aggregate energy usage for lighting devices.
GET server.com / api / measures / energy?category = lighting&location = "Living Room" – returns aggregate energy usage for lighting devices in the living room location.
GET server.com / api / measures / energy?category = lighting&location = "Living Room"&startDate=1 – returns up to 100 aggregate energy usage historical values for lighting devices in the living room location.

5

ModBus

5.1 Introduction

Many protocols have been designed for the needs of industrial automation and metering. These protocols generally use simple query/response models and allow for extremely simple implementations. Many protocols derived from the frame formats defined by IEC 870-5 such as:

- T101 (IEC 870-5-101) that was generated by the IEC TC57 for electric utility communication between master stations and remote terminal units, it is also based on the IEC-870-5-x link layer, using frame format FT 1.2.
- DNP 3.0, a protocol originally designed by Westronic, Inc. that was released into the public domain in 1993, based on the IEC-870-5-x link layer with a few modifications (e.g., use of FT3 frames for asynchronous, rather than synchronous, communication, inclusion of both source and destination addresses).
- M-Bus (see Section 9.3)
- Profibus, a fieldbus initially designed by Siemens and later standardized as IEC 61 158 ("Digital Data Communication for Measurement and control, Fieldbus for use in industrial control systems" for versions DP-V0, DP-V1 and DP-V2) and IEC 61 784 (Communication Profile Family DPF3). The protocols user's association website is http://www.profibus.com/.

Other protocols developed independently into *de-facto* standards, such as ModBus, a very common protocol that is used in many industrial and HVAC installations.

The Internet of Things: Key Applications and Protocols, First Edition.
Olivier Hersent, David Boswarthick and Omar Elloumi.
© 2012 John Wiley & Sons, Ltd. Published 2012 by John Wiley & Sons, Ltd.

5.2   ModBus Standardization

ModBus is a trademark of Modicon inc (Schneider Electric group), which also maintains the standard. The ModBus standard specification over a serial line can be found at http://www.modbus.org/docs/Modbus_over_serial_line_V1_02.pdf.

ModBus is an application layer messaging protocol that provides client/server communication between devices connected on different types of buses or networks. Because of its simplicity, ModBus has become one of the *de-facto* standards for industrial serial-message-based communications since 1979.

ModBus typically runs on top of RS 232, RS 442 point to point or RS 485 point to multipoint links. The ModBus/TCP specification, published in 1999 defines an IP-based link layer for ModBus frames.

ModBus devices communicate using a master-slave model: one device, the master, can initiate transactions (called *queries*), which can address individual slaves or be broadcast to all slaves. The slaves take action as specified by the query, or return the requested data to the master.

5.3   ModBus Message Framing and Transmission Modes

The transmission mode defines the framing and bit encoding of the messages to be transmitted on the ModBus network. In a given ModBus network, all nodes must use the same mode and serial parameters:

- In the *ASCII Transmission Mode*, each byte is encoded on the serial link as 2 ASCII characters. Each ASCII character is sent separately as 1 start bit, 7 data bits, zero or one parity bit, one or two stop bits. The message is framed by a starting ":" ASCII byte, and ends with a "CR-LF" byte sequence (see Figure 5.1).
- In the *RTU* (remote terminal unit) *transmission mode*, the message is transmitted in a continuous stream. Each 8-bit byte is framed by 1 start bit, 8 data bits, zero or one parity bit, one or two stop bits. The message itself starts after a silent period of at least 3.5 character times.

ModBus Addresses: ModBus messages begin by the target 8-bit address that can take any decimal value between 1 and 247. 0 is used for broadcasts. The address field of the message frame contains two characters in ASCII mode, or 8 bits in RTU Mode. Each query contains the address of a specific slave. When it responds, the slave includes its own address in the message.

ModBus Functions: The function code field contains two characters in ASCII mode, and 8 bits in RTU mode, which can take any decimal value between 1 and 255 and are selected based on the device application profile. Some example functions are listed:

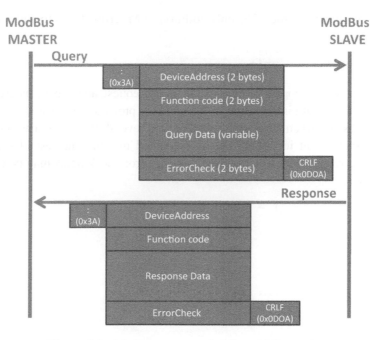

Figure 5.1 ModBus message framing (ASCII mode).

- 0x02: Read Input Status. Parameters: starting register address, and number of consecutive addresses to read. Response data: 1 bit per input read.
- 0x11: Report Slave ID. Parameters: none. Response data: slave ID, run indicator, device specific data.

ModBus Data Field: The data field provides the application level information, as required by the ModBus function. When a given ModBus function requires variable size data, the data field begins with the "byte count" of the data.

5.4 ModBus/TCP

The ModBus/TCP specification can be found at http://www.eecs.umich.edu/~modbus/documents/Open_ModbusTCP_Standard.doc

ModBus/TCP provides TCP/IP access to the ModBus functionality. Each ModBus Request/response is sent over a TCP connection established between the master and the slave, using well-known port 502. The TCP connection may be reused for several query/response exchanges.

The byte content of the ModBus request and response frames (i.e. without framing start-stop-parity bits specific to the serial physical layer) is simply transported over the TCP connection, in big indian order. The only addition of ModBusTCP is to add a seven-byte message prefix:

```
ref ref 00 00 00 len unit
```

The ref bytes are simply copied by the slave from the request, and may be used as a handle by the master. The length information in the message prefix allows proper reassembly of the ModBus message when it has been segmented in several IP packets. The slave address has been renamed "unit identifier" and is contained in unit. The rest of the message conforms to the regular ModBus structure, but the error check fields may be omitted for obvious reasons.

6

KNX

6.1 The Konnex/KNX Association

The Konnex (or KNX) Association was set up in 1999 on the merger between three former European associations promoting intelligent homes and buildings:

- Batibus Club International (BCI France) promoting the Batibus system;
- The European Installation Bus Association (EIBA) promoting the EIB system;
- European Home Systems Association (Holland) promoting the EHS system.

The goal of the KNX Association was to define and offer certification services for the KNX open standard, while offering legacy support and certification cervices for Batibus, EIB[1] and EHS. Membership is limited to manufacturers, there are over 200 members from 22 countries as of 2010, including ABB, Agilent, Bosch, Electrolux, Hager, Legrand, Merten, Moeller, Schneider, Siemens and many more leading vendors of home and building automation equipment.

KNX technology is royalty free for KNX members.

6.2 Standardization

In order to standardize the specifications, the KNX association cooperates with CENELEC TC 205. The KNX protocol has become an international standard in Europe as EN 50 090 (media and management procedures), EN 13 321-1 (media) and EN 13 321-2 (CEN, KNXnet/IP).

---

[1] EIB is backward compatible to KNX, most devices can be labeled both with the KNX as well as the EIB logo.

The Internet of Things: Key Applications and Protocols, First Edition.
Olivier Hersent, David Boswarthick and Omar Elloumi.
© 2012 John Wiley & Sons, Ltd. Published 2012 by John Wiley & Sons, Ltd.

At an international level, KNX is standardized by ISO and IEC (ISO/IEC 14 543-3). KNX is in prestandard stage in China as GB/Z 20 965.

There are several versions of KNX, which are all backwards compatible. The current version is 2.0 (since August 2009).

The overall specification counts over 6000 pages, divided into 10 volumes.[2] Individual specification documents can be purchased from the KNX association. The KonCert group manages KNX certification and testing.

Gateway specifications exist between BACnet (ISO 484 Annex 5 H.5 mapping KNX and BACnet, see also Chapter 3), DALI (lighting control) and KNX.

6.3   KNX Technology Overview

The overall KNX architecture is documented in Vol 3, part 3/1. The KNX architecture is decentralized: nodes can interact with other nodes without the need for a central controller.

The protocol stack uses the OSI model with a null session and presentation layer. It is based on the original work of EIB, which is therefore backward compatible to KNX.

KNX standardizes the protocol, but also the data model (EN 50 090-3-3, KNX volume 3/7) for basic types (integer and float values, percentage) and common device functions such as switching, dimming, blinds control, HVAC and so on . . .

6.3.1   Physical Layer

The physical layer of KNX is specified in Vol 3, Chapter 3/3/1 of the specifications. KNX can use a variety of physical layers

- **TP1[3]: Twisted pair** (Chapter 3/2/2). TP was the first physical layer that was defined as part of EIB, and is still the dominant physical layer used in KNX deployments. The TP bus provides both power and communication, using inductive coupling (Figure 6.2). A twisted-pair installation is made of lines, each line is composed of up to 4 line segments interconnected by repeaters, and each segment interconnects up to 64 devices. Lines are interconnected by line couplers (LC). The line couplers interconnect to the KNX backbone via a backbone controller (BbC), and the devices that can be accessed via a given BbC are part of the same KNX area (or zone). Line couplers and backbone controllers act as routers, that is, filter the messages that they relay based on

---

[2] Vol 1 Primer (deprecated), Vol 2 Cookbook (deprecated), Vol 3 System specifications, Vol 4 Hardware requirements (link to relevant standards, e.g., safety and environmental requirement), Vol 5 Certification manual, Vol 6 profiles, Vol 7 application descriptions (actual functional profiles), Vol 8 conformance testing, Vol 9 Basic and system Components (physical couplers, bus interface modules and couplers), Vol 10 Specific standards (Extended Tag format).

[3] There was a TP0 defined, which was deprecated by KNX.

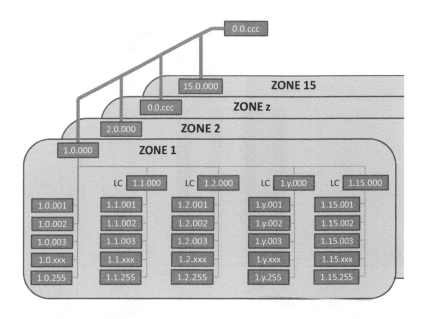

Figure 6.1   KNX network topology.

the destination address and the domain id (when present). The address space allows up to 15 areas (Figure 6.1), each with 15 lines, and a KNX TP installation can manage a maximum of 61 249 devices.

The physical layer uses a CSMA/CA medium access control using inductive coupling (Figure 6.2), and performs error detection (horizontal or vertical parity checks with acknowledgements (Ack, Nack, busy). The bus provides a 21 to 29 V power supply, and low-power KNX nodes may draw power from the TP line (typically up to 150 mW).

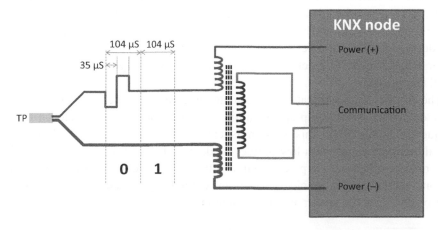

Figure 6.2   KNX TP node, modulation and inductive coupling.

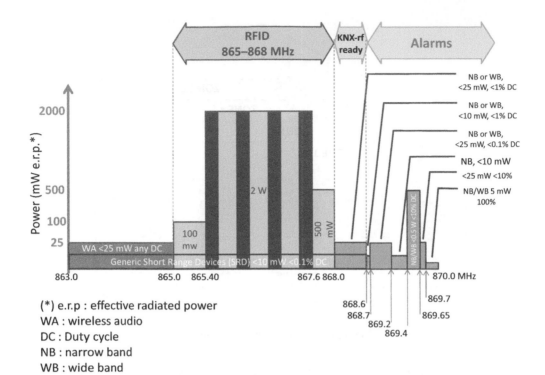

Figure 6.3   Usage of the 863–870 MHz band in Europe.

The transmission begins with a start bit (0), followed by a application octet, a parity bit, a stop bit (1) and a mandatory pause (11). The theoretical throughput is 9600 bps.

– **PL110 Over PLC**[4] (Chapter 3/2/3). PL110 uses a FSK modulation scheme and was also part of the original EIB specification. Each PLC line can have up to 64 devices. Since PLC is inherently a broadcast open media, the separation of domains (the portion of the KNX network logical topology over which the data signals of one physical layer type propagate) is ensured by a 48-bit domain address, in addition of the zone/line/node Id address (see TP1 for a description of these addresses).

– **Over RF** (Chapter 3/2/5 defined in 2001). This physical layer uses the 868-870 MHz band (Figure 6.3).

The KNX-rf 1.1 specification was updated in 2010, introducing a "push button" and easy controller mode setup specification, and using a 1% duty cycle on the center frequency 868.3 MHz: this version is called "KNX-rf ready". It allows bidirectional communication with low duty cycle devices by sending a 4.8-ms preamble for transmissions. Devices are preconfigured with group addresses for multicast communication, and unicast communication uses an "extended group address" composed of the group address of the sender and its serial number.

---

[4] There was a PLC 132, which was removed from the specification.

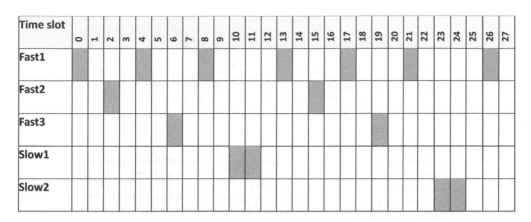

Figure 6.4 KNX-rf multireceiver scanning.

A further KNX-rf update is expected in 2011 called "KNX-rf multi" that introduces 3 RF "fast channels" and 2 RF "slow channels" (for battery-powered devices and energy-harvesting products) as well as fast acknowledge services. Fast channel 1 would use the existing KNX 1.1 center frequency (868.3 MHz) and a duty cycle of 1%, it is the configuration channel and the default call channel. Fast channel 2 would use a center frequency of 868.95 MHz and a duty cycle of 0-1%, Fast channel 3 (optional, coexisting with slow channel 1) would use a center frequency of 869.85 MHz and a duty cycle up to 100%. Slow channels use a preamble length of 500 ms and duty cycle of 10%. Slow channel 2 will use a center frequency of 869.525 MHz. KNX-rf "multi" products will not be compatible with KNX-rf 1.1. products, but will be compatible with KNX ready products. Because of the preamble, a KNX-rf multireceiver can scan both fast and slow channels (Figure 6.4).

- **Over IP** (Chapter 3/2/6). KNX may use IP as a native communication medium. KNXnet/IP uses binary or XML encoded PDUs to emulate bus communication. KNX/IP defines a tunneling mechanism over IP, using datalink layer binary PDUs.

6.3.2 Data Link and Routing Layers, Addressing

The data link layer is specified in Vol 6, Chapter 3/3/2, and the routing layer is specified in Chapter 3/3/3.

KNX nodes exchange telegrams, formatted as in Figure 6.5. Telegrams are transmitted octet by octet on the physical layer. Each telegram is acknowledged by the recipient node after a mandatory pause (equivalent to 11 bits), and retransmitted, if needed, up to 3 times. On TP, a 9 octet command will be transmitted and acknowledged (when unicast) in about 15 ms.

Each KNX node has a unique 2 byte address, used mainly for configuration purposes and as the source of telegrams. The source address is encoded on 16 bits as an area identifier (4 bits), a line identifier (4 bits) and a device number (8 bits). The source address

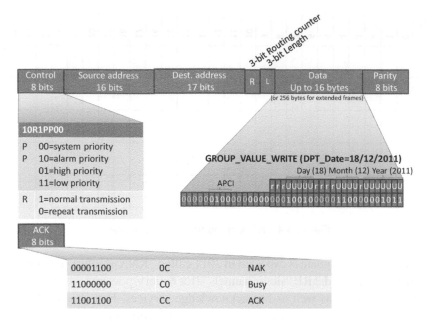

Figure 6.5   KNX telegram format.

appearing in a telegram is always a physical address. This address is configured during the installation process, typically by pressing a key on the device that causes it to enter a signaled state that allows the KNX ETS installation tool to assign an address to the device. Device number 0 is reserved for line couplers and backbone couplers.

The destination address may be a logical group address (bit $17 = 1$) or a physical address (bit $17 = 0$). The group address may be formatted on 2 levels (4-bit main group ID, 11-bit secondary group ID) or 3 levels (4-bit main group ID, 3-bit auxiliary group ID, 8-bit secondary group ID). The group destination address is the abstraction for a command wire: a KNX switch "connects" to a KNX lamp controller by sending a message to the group that is configured as the lamp controller input, the message acts as a data container for values that the sending nodes wants to share with the receiving node(s). Such group bindings are defined during the configuration of KNX nodes. Special group address 0000 h addresses all KNX nodes.

The KNX installation has a 64 k addressing space. The individual address and all group addresses of a device are stored in the device "address table".

KNX supports 4 priority levels encoded as specific values of the control bits (see Figure 6.5). In the case of collisions among same priority nodes, the node with the lowest physical address transmits first. On the TP1 physical layer, this priority mechanism derives from the fact that "0" is dominant when a simultaneous 0/1 transmission collision occurs.

The 3-bit routing counter implements a "hop count": a line coupler decides if a telegram can be transmitted to the other side based on the remaining hop count. The number of hops is normally limited to 6.

An 8-bit parity field secures the KNX telegram, it is based on a vertical parity scheme: bit i of the security field is set to 1 if the octet-wise sum of bits i of the previous telegram data is an even number.

6.3.3   Transport Layer

The KNX transport layer is specified in Volume 3, Chapter 3/4.

The transport layer (TL) provides a connection-oriented peer to peer communication service, providing a connect and disconnect primitive, a TL-acknowledgment, sequence counter and time-out management (typically 6 s for the configuration mode).

The transport layer also removes the source of the message before calling the application layer, and therefore the behavior of actuators never depends on the source of messages.

6.3.4   Application Layer

The KNX application layer is defined in Vol 3, Section 3/3/7. The application layer defines the group objects, and the exchange of group-object values via service requests, for instance "group value write" illustrated in Figure 6.5.

The application layer also defines the "property value write" service, which is used to set values and configuration parameters to KNX device interface objects.

6.3.5   KNX Devices, Functional Blocks and Interworking

Volume 3, part 7 is dedicated to interworking. The overall interworking model is outlined in 3/7/1. Chapter 3/4 defines the application environment, and Chapter 3/4/1 defines the application interface layer, including data representation models for group objects and interface objects. Such as data structures or flags (e.g., transmission allowed, write allowed, data has been written by the bus . . .).

KNX devices are defined by the functional blocks that they support. A functional block is a logical grouping of inputs, outputs and parameters that are useful to perform a certain function. For instance the "sunblind actuator basic" is one of the functional blocks defined by KNX, which specifies:

- A list of mandatory and optional inputs (Move UpDown, StopStep UpDown, Set Absolute Position blinds Percentage, WindAlarm . . .).
- A list of mandatory and optional outputs (Info Move UpDown, Current Absolute Position Blinds Length, . . .).
- A list of mandatory and optional parameters (Reversion Pause time, Move Up/Down time, Preset Slat angle, . . .).

Inputs, outputs and parameters are specified by their datapoint types (see Section 6.3.5.3, and a specific functional interpretation of the Datapoint values in the specific context of the functional block. Inputs, outputs and parameters may be published as properties or group objects.

KNX volume 6 provides an extensive library of functional blocks (over 146 as of 2011) such as dimming controllers, room temperature controllers, schedulers, system clock, alarm, and so on.

6.3.5.1   Group Objects

The process information of KNX functional blocks (input, output or parameter) may be published as a group object (GO). A group object may be read or written over the bus via dedicated multicast service primitives. The KNX specification of each functional block defines which of the inputs, outputs and parameters may, or must, be published as a group object.

The type of the group object is described by a datapoint (see Section 6.3.6.). A group object that sends its value may be configured with one and only one destination group address, but a group object may listen to several group addresses. The target and monitored group addresses are configured in the KNX device *Address Table*. All group objects linked to the same group address must be of the same datapoint.

For example, if the on/off group object of a presence sensor (an output) is assigned group address 1/1, and the on/off group object of a light controller (an input) is also assigned group address 1/1, then the presence sensor will control the light controller.

6.3.5.2   Interface Objects

Interface objects (IO) store certain properties of the device, mostly parameters (Figure 6.6). A node can have up to 256 interface objects. The type of each interface object is identified

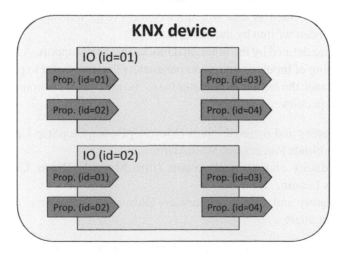

Figure 6.6   KNX interface objects and properties.

by a 16-bit identifier, the type of property is given by an 8-bit identifier. Chapter 3/7/3 contains the standard identifier tables.

The KNX application layer provides unicast primitives to read or set property values, using messages addressed to the physical address of the device.

The value of a group object can be reflected in the value of a property, in order to make it possible to read or set a property via group communication.

6.3.5.3 KNX Datapoints

Chapter 3/7/2 specifies the datapoint types. KNX provides an extensive library of datapoints (over 350 as of 2011), which are used to express the properties of KNX devices, and parameters of commands sent over the network (see Figure 6.5). Datapoints are defined by:

- Their data type (format and encoding);
- Their dimension (range and unit).

Each datapoint type is identified by a 16-bit main number.16-bit subnumber identifier. The main number identifies the format and encoding, the subnumber identifies the range and unit. Subnumbers are allocated by the KNX association based on the application domain: 0 to 99 for common use range and units, 100 to 499 for HVAC applications, 500 to 599 for load management, 600 to 999 for lighting, 1000 to 1999 for system applications. Subnumbers greater than 60 000 are used by manufacturer-specific extensions.

KNX has defined its own syntax to define Datapoints types, with a letter representing the data type of each field (e.g., unsigned value, see Figure 6.7), and a subscript number indicating the number of bits used to encode the data type. For instance, U_8 is an unsigned

A	Character
A[n]	String of n characters
B	Boolean/Bit set
C	Control
E	Exponent
F	Floating point value
M	Mantissa
N	eNumeration
R	Reserved bit or field
S	Sign
U	Unsigned value
V	2's Complement signed value
Z_8	Standardized Status/Command B8. Encoding as in DPT_StatusGen

Figure 6.7   KNX Datapoint field definition symbols.

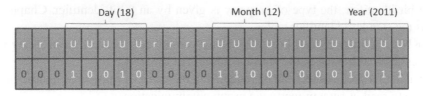

Figure 6.8 DPT_Date KNX datapoint.

number field encoded over 8 bits. DPT_Date is encoded as $r_3N_5r_4N_4r_1U_7$, and illustrated on Figure 6.8.

For metering applications, KNX has defined a number of datapoints aimed at tunneling M-Bus addresses and metering values, in order to facilitate interworking. See Section 9.3 for more details.

The library of datapoint types may be downloaded from the KNX Association's website for free.

Type information is used mainly at configuration time: it is not transmitted for better performance and to avoid imposing unnecessary restrictions on the combinations of devices.

6.4   Device Configuration

Device configuration uses mainly point to point (unicast) telegrams. There are three options to configure a node:

- In the system mode or "S-mode", the management configuration tool runs on a PC (using KNX ETS™ software, which stores configuration data as XML schema).
- In easy mode or "E-mode", several strategies are employed to avoid the use of a PC to configure the network. An embedded "master controller" may be activated to search partner devices, and connects to further devices one by one (identify/discover device). Group bindings may then be configured via a controller menu, or in "push button" mode. In the "push button" mode, links are configured one by one by first activating the actuator ("push-button"), then the sensor to be enrolled.
- Devices may also be preconfigured using a logical tag (extended possibility specified in Volume 10). The devices are preconfigured and use a specific framing format (EFF extended frame format), in additional of the standard format. This mode introduces semantical and geographical zoning tags. Configuration tuning is performed with ETS.

7

ZigBee

7.1 Development of the Standard

The 802.15.4 standard provides a physical and link layer technology optimized for low bitrate, low duty cycle applications. However, in practice sensor and control applications also need a mesh networking layer, and a standard syntax for application layer messages. In 2002, several companies decided to form the ZigBee alliance to build the missing standard layers that would be required to enable a multivendor mesh network on top of 802.15.4 radio links.

In 2008, the ZigBee alliance counted more than 200 members:

- *Promoter* members get early access to, contribute to and vote on the specifications of the alliance. They can veto decisions made by other participants in the alliance and get special marketing exposure in ZigBee events. New candidates for the promoter status must get co-opted by existing promoter members.
- *Participant* members have the same contribution and voting rights as promoters, but without veto rights.
- *Adopters* also get early access at the specifications, but can contribute only to the application profile working groups, and do not have voting rights.

The ZigBee alliance regularly organizes interop events, called ZigFests, and organize a developers conference twice a year. In order to ensure interoperability across vendors, the use of the ZigBee Compliant Platform (ZCP) certification and logo is reserved for products passing the ZigBee test suite, which includes interoperability tests with the "Golden units" (stacks from four reference implementations: Freescale, Texas Instruments, Ember, and Integration).

The deployment of many telecom standards either failed or was slowed down by multiple patent claims, many of which were not disclosed during the design phase of the

The Internet of Things: Key Applications and Protocols, First Edition.
Olivier Hersent, David Boswarthick and Omar Elloumi.
© 2012 John Wiley & Sons, Ltd. Published 2012 by John Wiley & Sons, Ltd.

standard. While the ZigBee alliance can do nothing against potential patent claims coming from nonmembers, it did verify that no technology included in the standard was subject of a known patent. In addition, every new member of the ZigBee alliance must sign a disclosure statement regarding patents that could potentially apply to ZigBee technology.

There are several versions of ZigBee. The current versions of ZigBee are ZigBee 2006/2007 (stack profile 0x01, ZigBee 2007 adds optional frequency agility and fragmentation), and ZigBee Pro (stack profile 0x02) that adds support for more nodes and more hops through source routing (it does not support tree routing), multicasting, symmetric links and a high security level. There was a ZigBee 2004 version, which is now deprecated.

7.2   ZigBee Architecture

7.2.1   ZigBee and 802.15.4

ZigBee sits on top of 802.15.4 physical (PHY) and medium-access control (MAC) layers, which provide the functionality of the OSI physical and link layers.

So far ZigBee uses only the 2003 version of 802.15.4. All existing ZigBee commercial devices use the 2.4 GHz S-Band as the 2003 version of 802.15.4 does not allow sufficient bandwidth on other frequencies. The 2006 version adds improved data-transfer rates for 868 MHz and 900 MHz but is not yet part of the ZigBee specification.

802.15.4 offers 16 channels on the 2.4 GHz, numbered 11 to 26. ZigBee uses only the nonbeacon-enabled mode of 802.15.4, therefore all nodes use CSMA/CA to access the network, and there is no option to reserve bandwidth or to access the network deterministically. ZigBee restricts PAN IDs to the 0x0000 – 0x3FFF range, a subset of the 802.15.4 PAN ID range (0x0000-0xFFFE).

All unicast ZigBee commands request a hop by hop acknowledge (optional in 802.15.4), except for broadcast messages.

7.2.2   ZigBee Protocol Layers

The ZigBee network layer provides the functionality of the OSI network layer, adding the missing mesh routing protocol to 802.15.4. It also encapsulates the network formation primitives of the 802.15.4 MAC layer (network forming and joining).

The rest of the ZigBee protocol layers (Figure 7.1) do not follow the OSI model:

– The **Application Support Sublayer (APS)** layer has several functions:
  • Multiplexing/demultiplexing: it forwards the network layer messages to the appropriate application objects, according to their endpoint ID (each application is allocated an endpoint ID).
  • Binding: the APS layer maintains the local binding table, that is, records remote nodes and endpoints which have registered to receive messages from a local endpoint.
  • 64-bit IEEE to 16-bit ZigBee network node address mapping.

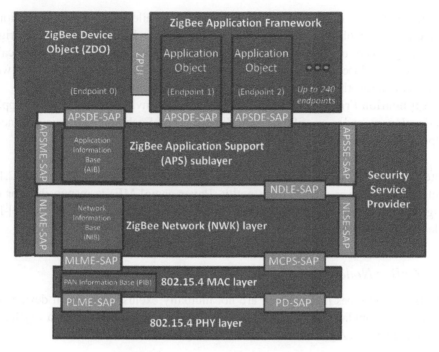

Figure 7.1 ZigBee architecture overview.

- Management of end to end acknowledgements. The application layer supports acknowledgements independently of the link layer acknowledgements of 802.14.4. The APS manages retries and duplicate filtering as required, simplifying application programming.
- Fragmentation.

Also, as part of the application support sublayer management entity, or APSME:

- Group addressing: the APSME allows to configure the group membership tables of each endpoint ID, and forwards messages addressed to a group ID to the application objects with relevant endpoint IDs.
- Security: management of keys.
– The **ZigBee Device Object (ZDO)** layer is a specific application running on endpoint 0, designed to manage the state of the ZigBee node. The ZDO application implements the interfaces defined by the ZigBee device profile (ZDP, application profile ID 0x0000). These primitives encapsulate the 802.15.4 network formation primitives of the ZigBee network layer (node discovery, network joining), as well as additional primitives supporting the concept of binding (see Section 7.5.2.2).
– The **ZigBee Cluster Library (ZCL)** was a late addition to ZigBee, specified in a separate document. It consists in a library of interface specifications (cluster commands and attributes) that can be used in public and private application profiles. It is now

considered as one of the key assets of ZigBee: while the ongoing evolution of ZigBee towards a 6LoWPAN-based networking layer is likely to replace the original networking layers of ZigBee, the ZCL is likely to remain the "lingua franca" of application developers. One important addition of the ZCL is the group cluster, which provides the network interface for group formation and management.

– The **Application Framework** layer provides the API environment of ZigBee application developers, and is specific of each ZigBee stack. Each application is assigned an Endpoint ID.

The interfaces of ZigBee layers are called "service access points" (SAP), as in 802.15.4. One interface, the layer management entity ([layer name]-ME) is responsible for configuring internal data of the layer. Another interface, the data entity ([layer name]-DE), provides the data send/receive and other nonmanagement primitives.

7.2.3   ZigBee Node Types

The ZigBee node types listed below are not mutually exclusive. A given device could implement some application locally (e.g., a ZigBee power plug) acting as a ZigBee End Device, and also be a ZigBee router and even a ZigBee coordinator.

– **ZigBee End-Device (ZED):** this node type corresponds to the 802.15.4 reduced function device. It is a node with a low duty cycle (i.e. usually in a sleep state and not permanently listening), designed for battery operation. ZEDs must join a network through a router node, which is their parent.
– **ZigBee router (ZR):** this node type corresponds to the 802.15.4 full function device (FFD). ZigBee routers are permanently listening devices that act as packet routers, once they have joined an existing ZigBee network.
– **ZigBee Coordinator (ZC):** this node type corresponds to a 802.15.4 full function device (FFD) having a capability to form a network and become a 802.15.4 PAN coordinator. ZigBee coordinators can form a network, or join an existing network (in which case they become simple ZigBee routers). In nonbeacon-enabled 802.15.4 networks, coordinators are permanently listening devices that act as routers, and send beacons only when requested by a broadcast beacon request command.

The ZigBee coordinator also contains the trust center, which is responsible for admission of new nodes on the network and management of security keys (see Section 7.7).

7.3   Association

7.3.1   Forming a Network

When forming a network, a ZigBee coordinator first performs an active scan (it sends beacon requests) on all channels defined in its configuration files. It then selects the channels with the fewer networks, and if there is a tie performs a passive scan to determine the quietest channel. It finally broadcasts a 802.15.4 beacon for the selected PAN ID on the selected radio channel, then remains silent (or repeats the beacon periodically, depending on the implementation). Depending on the configuration of the stack, the scan duration on each channel can range from 31 ms to several minutes, so the network-forming process can take significant time. If there are any ZigBee routers associated to the network, they will typically repeat the beacon with an offset in time relative to their parent's beacon (an extension of 802.15.4:2003).

The ZigBee specification allocates range 0x0000 to 0x3FFF for PAN IDs (a subset of the range defined by 802.15.4: 0x0000 to 0xFFFE). The PAN-ID should be unique for a given channel for networks not capable of dynamic channel change (ZigBee 2006), and unique on all channels if channel agility is enabled (ZigBee 2007, ZigBee PRO). A ZigBee coordinator beacon may also include an extended PAN ID (64 bit EPID), in addition of the 16-bit 802.15.4 PAN identifier, in order to facilitate vendor specific network selection for joining nodes. This EPID identifier is only used in the beacon frames and has no other uses, while the 16-bit 802.15.4 PAN identifier is always used for joining and addressing purposes.

7.3.2   Joining a Parent Node in a Network Using 802.15.4 Association

ZigBee devices that are not yet associated either capture by chance the beacon, or try to locate a network by broadcasting a 802.15.4 beacon request on each of the 16 radio channels (active scan, see Figure 7.2), unless the radio channel has been preconfigured or determined in the application profile. If a coordinator has formed a network on one of those channels, it responds to the beacon request by broadcasting a 802.15.4 beacon, which specifies the 16-bit PAN ID of the network, the address of the coordinator in short 16-bit format or extended 64-bit format, and optionally an extended PAN ID (EPID). Any ZigBee router that has already joined the network will also respond with a beacon if they hear the beacon request.

The ZigBee payload of the 802.15.4 beacon also contains the ZigBee stack profile supported by the network, a flag indicating whether the responding node has remaining capacity for routers or end devices joining as new children, and the device depth of the sending device, that is, its level in the parent/child tree rooted at the coordinator.

Once it has discovered the PAN ID of the network, its radio channel, and the address of a router or coordinator within radio reach, the new ZigBee node sends a standard 802.15.4 *association request* command to the address of the specific parent node it wants to join as a child node (0x01 profile nodes must join the node with the smallest device depth). The association request message uses the extended 64-bit address of the joining node as the source address. Devices may wish to join a specific PAN ID, or may use a special PAN ID value 0xFFFF to signal that they are willing to join any PAN ID.

The parent node acknowledges the command, and then if it accepts the association responds with a 802.15.4 association response command sent to the extended address of the device. The association response specifies the 16 bit short address that the device

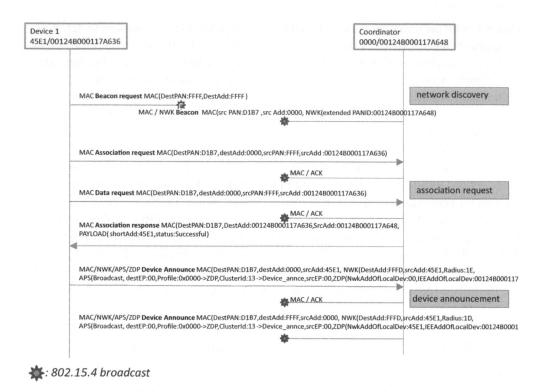

🔆: 802.15.4 broadcast

Figure 7.2   End device 1 joins the ZigBee network.

should use in the future (in order to save transmission time and therefore energy). The association response is acknowledged by the device. When the joining device is a sleeping node (RxOnIdle=false), the association response is not sent immediately, but stored until the sleeping nodes polls it using a "data request", as in the example of Figure 7.2.

Once associated, ZigBee devices usually send a data request command (now using the short 16-bit address assigned by the parent as source address) to its parent in order to receive any pending configuration data. After waiting for a response, battery-powered ZigBee end devices go to sleep until the next scheduled wake-up time or interrupt.

In the example of Figure 7.2, the joining device is within radio range of the coordinator. In a more general case, broadcast and unicast messages will be relayed by one or more ZigBee routers, and a new joining device may join the network from any location accessible through mesh networking.

ZigBee joining, at the lowest level of security, only uses the "permit joining" flag of the beacon: nodes can join a network only when this flag is set in the beacon response. At the application level, most implementations allow administrators to "permit joining" for a limited amount of time, after which the network will not accept further joins. If the device is allowed to join and the nwkSecurityLevel parameter is set to 0x00, then the node becomes a new child of the parent node with relationship type 0x01 (child), otherwise set as type 0x05 (unauthenticated child). When security is enabled, interactions with the trust center (see Section 11.2.1) follow the unauthenticated joining process for key distribution.

In order to facilitate commissioning, nodes may implement the *commissioning ZDP cluster* (see Section 7.6) to preconfigure security material and other parameters, and reset the node. Some nodes may be set up to join any network with permit-join enabled, or may be preconfigured to join the well-known commissioning network with extended PAN ID 0x00f0c27710000000.

In theory, up to 31 100 nodes (9330 routers) can join a given network in stack profile 0x01, and over 64 000 nodes in stack profile 0x02.

7.3.3   Using NWK Rejoin

A device that loses connection to the network can attempt to rejoin using the ZigBee NWK layer rejoin command, which also triggers a beacon request. Since the NWK layer rejoin command use NWK layer security, the difference from a join based on 802.15.4 association is that no additional authentication step needs to be performed when security is enabled, and that nodes may rejoin any parent as long as it has available capacity, regardless of the status of the accept joining flag of the beacon. If it rejoins a different parent (e.g., because the original parent no longer responds), the node will be allocated a different short address, and must broadcast a device announce to the network in order to update bindings that may be configured in other nodes (see Figure 7.2).

After power cycles, most implementations do not immediately attempt an explicit rejoin in order to avoid network overloads, if they still have the address of their parent node and their own short address in nonvolatile memory. It is assumed that all nodes will restart in the same state as before the power cycle. An explicit rejoin is triggered only if the node fails to communicate with its parent. Such a procedure is often referred to as "silent rejoin". It is also the default procedure, in ZigBee Pro/2007, when the coordinator triggers a channel change (annex A).

7.4   The ZigBee Network Layer

The network layer is required for multihop routing of data packets in the mesh network, and is one of the key missing elements of 802.15.4. ZigBee uses the AODV public-domain mesh algorithm. The ZigBee network layer uses a specific data frame format, documented in Table 7.1, which is inserted at the beginning of the 802.15.4 payload.

7.4.1   Short-Address Allocation

ZigBee uses the 0x0000 – 0xFFF7 range for network node short addresses. The ZigBee coordinator uses short address 0x0000. The allocation of other network addresses, under control of the ZigBee Coordinator, depends on the routing technology in use:

ZigBee supports two address allocation modes:

- In stack profile 0x01, the network address depends on the position of the node in the tree. The distributed address assignment mechanism uses CSkip, a tree-based network address partition scheme designed to provide every potential parent with a subblock of network addresses. In addition to the default meshed routing, a tree-based routing can be used as a back-up (routers use the address allocation to decide whether to forward the packet to a parent or to a child).
- In stack profile 0x02 (ZigBee 2007, ZigBee Pro), a stochastic address assignment mechanism is used and ZigBee provides address-conflict detection and resolution mechanisms.

7.4.2   Network Layer Frame Format

The network layer PDU format is illustrated in Table 7.1, and is transported as 802.15.4 payload (see Chapter 1).

Table 7.1   The ZigBee network layer frame format

Field name	Size (octets)	Field details
Frame Control	2	--------------XX : Frame type (00 : network data)
		----------0010-- : Protocol version (always 0x02 for ZigBee 2006/2007/Pro)
		--------XX------ : Route discovery (0x01:enable)
		-------X-------- : Multicast (0 : unicast)
		------X--------- : Security (0 : disabled)
		-----X---------- : Source route (0 : not present)
		----X----------- : Destination IEEE address (0 : not specified)
		---X------------ : Source IEEE address (0 : not specified)
		000------------- : Reserved
Dest. Address	2 or 8	0xffff broadcast to all nodes including sleeping devices
		0xfffd broadcast to all awake devices (RxOnIdle = True)
		0xfffc broadcast only to routers, not to sleeping devices
Source address	2 or 8	
Radius	1	Maximum number of hops allowed for this packet
Sequence number	1	Rolling counter
Payload	Variable	APS data, or network layer commands

7.4.3   Packet Forwarding

At the network layer, ZigBee packets can be:

- *Unicast*: the message is sent to the 16-bit address of the destination node
- *Broadcast*: if broadcast address 0xFFFF is used, the message is sent to all network nodes. If broadcast address 0xFFFD is used, the message is sent to all nonsleeping nodes. If broadcast address 0xFFFC is used, the message is broadcast to routers only (including the ZigBee coordinator). A radius parameter adjusts the number of hops that each broadcast message may travel. The number of simultaneous broadcasts in a ZigBee network is limited by the size of the broadcast transaction table (BTT), which requires an entry for each broadcast in progress. The minimal size of the BTT is specified in ZigBee application profiles, for example, 9 for HA.
- *Multicast* that is, sent to a 16-bit group ID.

ZigBee unicast packets are always acknowledged hop by hop (this is optional in 802.15.4). Broadcast packets are not acknowledged and are usually retransmitted several times by the ZigBee stacks of the originating node and by ZigBee routers on the path. Note that broadcast messages and messages sent to group IDs are not always broadcast at the link layer level: since sleeping nodes do not receive such messages, after wake up they send a data request to their parent node, and the queued messages are sent as unicast messages specifically to the sleeping node (in which case they are acknowledged at the link-layer level). In most implementations, the messages are queued only 7 to 10 seconds in the parent node (ZigBee specifies 7 seconds) so sleeping nodes should wake up at least once every 6 seconds. However in practice several vendors manufacture sleeping nodes with wake up periods of up to 5 minutes ... in this case applications need to be prepared to resend commands every 6 seconds until the target node wakes up.

Typically, a ZigBee node forwards a packet in about 10 ms. The propagation of data packets through the meshed network is limited by the initial value of Radius, a hop counter decremented at each routing node.

Delivery of packets to sleeping nodes uses the IEEE "Data request" packet. Parent nodes buffer received packets for their sleeping children. When it wakes up, the sleeping child sends a IEEE "Data request" packet to its parent. If it has data pending, the parent sets a specific "more data" flag in the ACK response, then the pending data.

7.4.4   Routing Support Primitives

The network layer provides a number of command frames listed in Table 7.2.

The route request command enables a node to discover a route to the desired destination, and causes routers to update their routing tables. At the MAC level, the route request command is sent to the broadcast address (0xffff) and the current destination PAN ID.

The response, if any, is a route reply command that causes routers on the path to update their routing tables.

Table 7.2 ZigBee routing layer primitives

Command Frame Identifier	Command Name Reference
0x01	Route request
0x02	Route reply
0x03	Network Status
0x04	Leave
0x05	Route record
0x06	Rejoin request
0x07	Rejoin response
0x08	Link status
0x09	Network report
0x0a	Network update

7.4.5   Routing Algorithms

7.4.5.1   Broadcast, Groupcast, Multicast

At the 802.15.4 level, messages sent to multiple destinations are always broadcast. At the network layer, however, ZigBee offers more possibilities, depending on the destination address:

- 0xffff broadcast to all nodes including sleeping devices;
- 0xfffd broadcast to all awake devices (RxOnIdle = True);
- 0xfffc broadcast only to routers, not to end devices.

In order to avoid 802.15.4 collisions, broadcast packet are relayed after a random delay of about 100 ms and therefore propagate ten times more slowly than unicast messages. The radius parameter is decremented at each hop, so the broadcast propagation can be controlled with the initial radius value (see Table 7.1).

ZigBee also uses a broadcast transaction table (BTT) in order to avoid any looping of broadcast messages: each broadcast packet is uniquely identified by its source address and network sequence number. When relaying a broadcast message, routers keep a copy of this unique identifier for 9 s (broadcast timeout), and will drop any looped packet. If the BTT is full, all broadcast messages are dropped. Routers that do not hear all neighbor routers retransmit a broadcast message may retransmit the broadcast message, implementing a form of implicit acknowledge mechanism.

Groupcast and multicast are implemented by the APS layer:

- APS messages sent to group addresses are filtered by the APS layer of the receiving node, so that only endpoints (and all of them) of member nodes will receive the message. However, all nodes receive the message (destination set to 0xffff at the network layer)

- In ZigBee Pro, the radius is not decremented when the message is forwarded by a group member. This makes it possible to restrict the 802.15.4 broadcast propagation to group members only, allowing some slack (apsNonmemberRadius) in order to cope with disconnected groups. ZigBee Pro calls this "multicast".

7.4.5.2 Neighbor Routing

This mode is not formally documented in ZigBee, but most vendors use it. If a router R already knows that the destination of a packet is a neighbor router or a child device of R, it can send the packet directly to this node. ZigBee end devices, however, must always route outgoing packets to their parent.

7.4.5.3 Meshed Routing

This is the default routing model of ZigBee. It implements the advanced *ad-hoc* on-demand distance vectoring (AODV) algorithm.

The principle of AODV is illustrated on Figure 7.3.

Node A needs to set up a route to node D. It broadcasts a route request (see Table 7.3) to network address 0xfffc (routers only), which propagates through the network. Each ZigBee router that receives that message forwards it to its neighbors, adding their local estimation of quality of the link over which they received the route request to the path cost parameter of the route request. Note that the route we are discovering is in the A to D direction, while the path costs actually used are in the D to A direction. This is because the sender of the route request, which is broadcasting the message, cannot transmit different values of the link cost, therefore the receiving node needs to update the path cost. *ZigBee mesh routing assumes symmetrical link quality.*

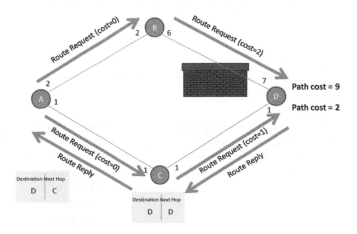

Figure 7.3   ZigBee mesh route discovery.

Table 7.3   Route request parameters

Command options	Route request identifier	Destination address	Path cost	Destination IEEE address
(1 octet)	(1 octet)	(2 octets)	(1 octet)	(0 or 8 octets)
xxx00xxx: not a many-to-one route request **xxx01xxx**: many-to-one route request and the sender supports a route record table **xxx10xxx**: many-to-one route request and the sender does not support a route record table **xxxxx1xx**: the command frame indicates the destination IEEE address, otherwise set to 0 **xxxxxx1x**: route request for a multicast group, and the destination address field contains the group ID, otherwise set to 0	Sequence number	Intended destination	Accumulator for path length as the command is propagated	Only if bit 5 of the command option is set to 1

Node D will wait for a while until it believes it has received all broadcast route requests, then computes the lowest path cost and responds by a unicast route reply along the best route that was just discovered (the parameters of the response are listed in Table 7.4). Router C creates a routing table entry for D, recording the next hop to reach D (in this simple case, D itself), and node A also creates a routing table entry for D, recording C as the next hop router.

Table 7.4   Route response parameters

Command options	Route request identifier	Originator address	Responder address	Path cost	Originator IEEE address	Destination IEEE address
(1 octet)	(1 octet)	(2 octets)	(2 octets)	(1 octet)	(0 or 8 octets) Address of the originator of the route request.	(0 or 8 octets) Address of the responder.

Node A will continue to use this route as long as it works, or until the applications requests calculation of a new route.

In ZigBee 2006 and 2007, routes are unidirectional: D will need to discover a route if it needs to send a packet to A. In ZigBee Pro, D would also store the route to A, assuming symmetry. ZigBee Pro routers "ping" each other every 15 s to make sure links are indeed bidirectional, and eliminate one-way links for both directions.

7.4.5.4 Tree-Based Routing

Tree-based routing is a back-up routing mechanism (when no route exists or can be discovered) that can be used only in ZigBee 2006/2007 networks which use Cskip-based address allocation. Tree-based routing simply uses the fact that routers know the blocks of addresses allocated to their router children, and to their children ZigBee end devices: if the destination address is one of those, the router forwards the packet to the appropriate child, otherwise it propagates the packet to its parent.

Because of the limited address space of 802.15.4 (64 K addresses), which limits the size of CSkip address blocks, tree-based routing is limited to 5 parent/child relationships, each node having a maximum of 20 children and 6 routers. This limits the number of nodes to 31 101 in ZigBee 2006/2007.

7.4.5.5 Source Routing

ZigBee mesh networking is limited by the size of the routing table of routers. It works and scales well in a peer to peer environment, but in environments where one node is a preferred communication source that frequently communicates to all other nodes, then this node and adjacent routers would need to store a routing table with as many entries as nodes. This situation happens with data concentrators, used for metering applications.

ZigBee Pro solves this problem by introducing source routing:

- The concentrator broadcasts a many to one route request (up to five hops), which enables routers along the path to record the shortest path to the concentrator.
- When a node first sends a packet to the concentrator, it first sends a route record message towards the concentrator, so that the concentrator will have a chance to learn the optimal path back towards the node (assuming symmetric links).
- The concentrator can now reach any node using source-routed messages (up to 5 intermediary routers can be specified). It still needs a lot of memory, but all routers in the ZigBee network can now work with very small routing tables.

7.5 The ZigBee APS Layer

The APS layer is responsible for management and support of local applications. It defines all the concepts that make it possible to develop and interconnect ZigBee applications: endpoints, groups, bindings, and so on.

7.5.1   Endpoints, Descriptors

A given ZigBee device may implement multiple applications at the same 802.15.4 address. Clearly some multiplexing mechanism is required to identify the source and destination application of a message. This multiplexing identifier is called *endpoint* in the ZigBee specification. Think of it as the equivalent of a port number in a TCP/IP network.

Each endpoint is further characterized by a simple descriptor. The simple descriptor contains the endpoint number, application profile ID, application device ID (a 16-bit number referring to a device definition, for example, a HA thermostat, used only for informative purposes since it contains no technical data), an application version ID, the list of input clusters and the list of output clusters.

The descriptor of an endpoint can be retrieved from any other node by using the ZDP simple descriptor request command (Simple_Desc_req, see also Section 7.8.1).

In addition to the simple descriptor, specific to each endpoint, ZigBee devices have additional descriptors applying to the whole node:

- A node type descriptor: capabilities of the node;
- A node power descriptor: node power characteristics;
- A complex descriptor (optional): Further information about the device descriptions, described as pairs of compressed XML tag and related field data;
- A user descriptor: User-definable descriptor.

Profiles can be discovered by using the ZigBee device profile request primitive, addressed to endpoint 0 (ZDO) of the device.

7.5.2   The APS Frame

APS data frames can be sent unicast (with or without application level end to end acknowledgment, in addition to the MAC level hop per hop acknowledgment), groupcast, multicast (ZigBee 2007 and ZigBee Pro), or broadcast. Groupcasts and broadcasts are both supported by network-level broadcasts, and are not acknowledged.

At the application level, ZigBee allows application developers to use 64-bit or 16-bit addresses, group addresses or indirect addressing in order to identify the destination node, for instance in the *APSDE-DATA.request*. In all cases, the ZigBee stack resolves that address to a 16-bit node address or to a group address before transmission.

This resolution mechanism uses the APS address map. This cache stores the mapping of 64-bit IEEE addresses to 16-bit ZigBee short network addresses. It is used, for instance, to resolve binding requests (which specify only a 64-bit IEEE address) to a 16-bit address. The maintenance of this table is performed by listening to broadcast device announce commands (e.g., when a device changes location and its 16-bit address changes, see Figure 7.2 for an example).

If a node does not have a cached route to the destination, it performs a route discovery using ZDP commands `IEEE address request` and `NWK address requests`. When a frame that required an end to end acknowledgment has not been acked after 3 retries (typically a retry every 1.5 s, this delay is adjustable in most stacks), another route discovery may be performed.

7.5.2.1 Groups

A group identifier (in the range 0x0000 to 0xFFFF) is an address that can be used at APS layer level to send a message to multiple ZigBee applications residing on other nodes (see Section 7.4.5.1). Any ZigBee node can belong to up to 16 groups. An application residing on a ZigBee node on endpoint E adds itself to a group G by calling the local AddGroupRequest APS primitive: this function adds endpoint E to the list of local members of group G.

Messages addressed to a group are broadcast at the ZigBee network layer (the ZigBee network frame destination address is 0xffff)[1]: ZigBee routers will forward a copy of the packet to any neighbor. Group messages are therefore received and processed by all nodes in the PAN network at the MAC layer, but the APS layer forwards the message only to the endpoints (individual applications residing on the node) that have registered to be members of the group ID.

The ZigBee HA profile recommends group addressing each time a message needs to be sent to more than 4 nodes.

7.5.2.2 Indirect Addressing, Binding

Bindings are one of the publish/subscribe models implemented in the ZigBee specification (together with attribute reporting, see Section 7.8).

Cluster C (see Section 7.8) on source endpoint E1 is bound to destination endpoint E2 (typically hosted by a different node) if it sends events related to its output cluster ID(s) to the corresponding input cluster ID(s) of E2. E1 can be bound to multiple target endpoints. Each binding is unidirectional and independent, if E1 is bound to E2, E2 may or may not be bound to E1 (it is a totally different binding).

The binding table can be managed locally through an API (APSME-BIND.request) or remotely via ZDP commands: The ZDP end-Device-Bind request is sent to endpoint 0 of the target node and specifies:

− The target endpoint of the binding;
− The source 64-bit IEEE address (the 16-bit ZigBee network address is resolved by the APS network address map);

---

[1] At the 802.15.4 level the ZigBee group messages might be unicast (if the sending node only has one neighbor) or broadcast.

– The source endpoint;
– The list of input clusters and output clusters of the source endpoint.

The local binding table lists, for each binding:

• The local source endpoint;
• The application layer destination address that can be a 802.15.4 address (64-bit format), or a group ID (16-bit);
• The destination endpoint if the destination is not a group address;
• A cluster ID.

A typical use case is that a device looks for another node in the network with capabilities corresponding to a match descriptor (supported application profile, cluster ID and direction, e.g., a lamp supporting the on/off cluster as an input). It then binds to that node using the End-Device-Bind command. In order to facilitate this configuration operation, bindings may be specified for groups. ZigBee devices that can initiate or process events have a button that places them in "identify mode" for about 10 s. Command AddGroupIfIdentifying can be broadcast and will automatically place the nodes in "Identify" mode in the group.

At the application level, the binding table can be used through the indirect addressing mode, for example, in the *APSDE-DATA.request* primitive. When indirect addressing is used, the destination address (node address or group address) is resolved using the local binding table, based on the endpoint ID of the sending application. Indirect addressing is very flexible as it allows external nodes to configure the routing of messages across ZigBee applications residing on different nodes (e.g., instruct a switch to send its on/off events to a ZigBee-controlled relay).

7.5.2.3 APS Frame Format

The APS frame format is outlined in Table 7.5.

The format of the application level payload depends on the value of the application profile identifier and the cluster identifier:

– Application Profile ID 0x0000: the payload format is defined by the ZigBee device profile (ZDP).
– Application Profile IDs 0x0000 to 0x7FFF are reserved for public application profiles, the payload format is defined by the ZigBee cluster library (ZCL)
– Application Profile IDs 0xBF00 to 0xFFFF are reserved for manufacturer specific profiles (MSP). The payload format is defined by the manufacturer but may also use the ZCL.

Table 7.5 The APS frame format

Field name	Bytes	Field details
Frame Control	1	------XX : Frame type (00 : APS data)
		----XX-- : Delivery mode (00 : unicast; 11 : group addressing)
		---X---- : Indirect address mode (0 : ignored)
		--X----- : Security (0 : none)
		-X------ : Ack (0 : not required)
		0------- : Reserved
Dest. Endpoint	1	16-bit destination address or group ID
Cluster identifier	2	0x0006: On/Off
Application Profile identifier	2	0x0104: HA
Source endpoint	1	
Counter	1	APS level counter
AF payload	Variable up to 80 bytes	APS service data unit (ASDU): a ZCL frame, ZDP frame or application-specific payload.

The ZDP and ZCL are described in the following Sections 7.6 and 7.8.

7.6 The ZigBee Device Object (ZDO) and the ZigBee Device Profile (ZDP)

The **ZigBee Device Object (ZDO)** layer is a specific application running on endpoint 0, designed to manage the state of the ZigBee node. The ZDO application implements the interfaces defined by the ZigBee device profile (ZDP, application profile ID 0x0000).

The clusters defined within the ZDP are similar to those defined in application-specific profiles, but unlike them, the clusters within the ZigBee device profile define capabilities supported in all ZigBee devices. All ZDP client-side transmission of cluster primitives is optional.

The ZDO implements a number of configuration attributes (e.g., the various node descriptors), as well as a number of local APIs and network primitives.

7.6.1 ZDP Device and Service Discovery Services (Mandatory)

These primitives support the discovery of nodes based on some of their characteristics. Since sleeping devices are not capable of receiving such requests, a cache mechanism is provided. The discovery cache is a database of nodes that registered to this cache after a find node cache request, and stores cached descriptor data (stored using Node_Desc_store_req).

ZDP primitives with mandatory server-side processing:

- NWK_addr_req (NWK_addr_rsp): finding a network address from a given IEEE address.
- IEEE_addr_req (IEEE_addr_rsp): finding an IEEE address from a given NWK address.
- Node_Desc_req, Power_Desc_req, Simple_Desc_req (and corresponding responses): finding the node/power/simple descriptor of a device from a given NWK address.
- Active_EP_req (Active_EP_rsp): acquire a list of endpoints on a remote device with simple descriptors.
- Match_Desc_req (Match_Desc_rsp): finding a list of devices with matching profile ids and cluster IDs.
- Device_annce.

ZDP primitives with optional server-side processing:

- Complex_Desc_req, User_Desc_req, Discovery_Cache_req, User_Desc_set, System_Server_Discover_req, Discovery_store_req, Node_Desc_store_req, Power_Desc_store_req, Active_EP_store_req, Simple_Desc_store_req, Remove_node_cache_req, Find_node_cache_req, Extended_Simple_Desc_req, Extended_Active_EP_req ... and corresponding responses.

7.6.2 ZDP Network Management Services (Mandatory)

These services implement the network features corresponding to the node type (coordinator, router or end device), for example, managing network scan procedures, interference detection, and so on. It provides the related interfaces for local applications. Remote nodes can also send a remote management command to permit or disallow joining on particular routers or to generally allow or disallow joining via the trust center.

Only the Mgmt_Permit_Joining_req/rsp server-side processing is mandatory, all other primitives are optional:

- Mgmt_NWK_Disc_req/ Mgmt_NWK_Disc_rsp (control network scanning).
- Mgmt_Lqi_req / Mgmt_Lqi_rsp (getting the neighbor list from a remote device).
- Mgmt_Rtg_req./ Mgmt_Rtg_rsp (getting the routing table from a neighbor device).
- Mgmt_Bind_req. Mgmt_Bind_rsp (getting the binding table from a neighbor device).
- Mgmt_Leave_req / Mgmt_Leave_rsp (request a remote device to leave the network).
- Mgmt_Direct_Join_req / Mgmt_Direct_Join_rsp (requesting that a remote device permit a device designated by DeviceAddress to join the network directly).
- Mgmt_Cache_req / Mgmt_Cache_rsp (allows to retrieve a list of ZigBee end devices registered with a primary discovery cache device).
- Mgmt_NWK_Update_req / Mgmt_NWK_Update_rsp (allows communication of updates to the network configuration parameters).

7.6.3   ZDP Binding Management Services (Optional)

These primitives enable binding management and maintenance of the bindings table (e.g., processes device replacement notifications). The concept of binding and indirect addressing is discussed in Section 7.5.2.2.

- End_Device_Bind_req / End_Device_Bind_res;
- Bind_req / Bind_res;
- Unbind_req / Unbind_res;
- Bind_Register_req / Bind_Register_res;
- Replace_Device_req/ Replace_Device_res;
- Store_Bkup_Bind_Entry_req/ Store_Bkup_Bind_Entry_res;
- Remove_Bkup_Bind_Entry_req/ Remove_Bkup_Bind_Entry_res;
- Backup_Bind_Table_req/ Backup_Bind_Table_res;
- Recover_Bind_Table_req / Recover_Bind_Table_res;
- Backup_Source_Bind_req / Backup_Source_Bind_res;
- Recover_Source_Bind_req / Recover_Source_Bind_res.

7.6.4   Group Management

Although one would expect group management over the network to be part of the core ZDP primitives, these were added later as part of the ZCL groups cluster (cluster ID 0x0004). See Section 7.8.

7.7   ZigBee Security

7.7.1   ZigBee and 802.15.4 Security

ZigBee networks can choose whether to enable security or not. Devices compliant with a public application profile must conform to their profile security settings.

ZigBee offers security services at two levels:

- Network (NWK) level security;
- Application (APS) level security.

None of these security services uses the MAC-level security defined by 802.15.4 (the 802.15.4 frame control field security bit is set to 0), which would encrypt the ZigBee NWK header that is required by ZigBee routing. However, ZigBee simply transposes the exact same mechanisms to the ZigBee network and application layer, and uses the

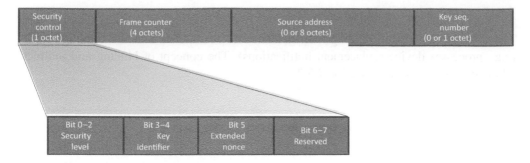

Figure 7.4 Auxiliary Security header format.

same encryption and hash algorithm (AES-CCM*), so encryption acceleration modules of 802.15.4 chips can still be used.

The ZigBee device object (ZDO) manages the security policies and configuration of a device.

7.7.1.1 NWK Level Security

ZigBee provides optional integrity protection and encryption, as illustrated in Figure 7.6. When network-level integrity protection or encryption is used, the security bit in the NWK control field is set to 1, indicating the presence of a security auxiliary frame header, and a message-integrity protection code (MIC).

The NWK security auxiliary header is composed of four subfields, illustrated in Figure 7.4.

In the security control subfield, the security level is set to the value of the MIB nwk security-level parameter, by default 0x05. This value specifies whether the frame is only integrity protected but not encrypted, or that it should also be encrypted, in which case the entire network payload is encrypted. Network security is applied as configured in the device network information base (NIB, nwkSecureAllFrames=TRUE to secure all frames) by default, but the application may override this setting frame by frame by specifying the SecurityEnable parameter of the NLDE-data.Request primitive.

The services provided by each security level are listed in Figure 7.5.

The MIC, as well as encryption, are computed using AES and CCS* (see Section 1.1), using the main or alternate network key.

7.7.1.2 Application Layer Security

The APS layer provides a number of security primitives that can be used by application developers:

- APSME-ESTABLISH-KEY to establish a link key with another ZigBee device using the SKKE protocol.
- APSME-TRANSPORT-KEY to transport security material from one device to another.
- APSME-UPDATE-DEVICE to notify the trust center when a device joins or leaves the network.
- APSME-REMOVE-DEVICE to instruct a router to remove a child from the network (used by the trust center).
- APSME-REQUEST-KEY to ask the trust center an application master key or the active network key.
- APSME-SWITCH-KEY used by the trust center to tell a device to switch to a new network key.
- APSME-AUTHENTICATE used by two devices to authenticate each other.

Security control	0x00	0x01	0x02	0x03	0x04	0x05	0x06	0x07
encryption	No				Yes			
MIC bit size	No MIC	32	64	128	No MIC	32	64	128

Figure 7.5 AUX header security control field values. The key identifier is set of 0x01 for the active network key (0x00 for link keys used by the APS layer).

In addition, the application-layer security provides its own optional integrity and encryption services, based on the network key or on a link-specific key (associated to the destination of the packet), under control of the application (TxOptions parameter).

Just like NWK security, the APS layer security will add an auxiliary security header (AUX header), and an integrity code, as illustrated on Figure 7.6. The difference is that the integrity protection scheme protects the APS header, auxiliary header and APS payload, and when encryption is used, only the APS payload is encrypted.

7.7.2   Key Types

- **Master Keys**
  - Application master keys are distributed by the trust center (via unsecured key transport) and used to set up link keys between two devices, and for mutual authentication of devices.
  - Trust center master keys are used to derive a link key for communication with the trust center.

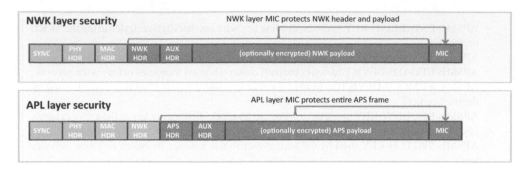

Figure 7.6   ZigBee NWK and APS security services.

- **Network keys** are used for network management. They are preconfigured or configured
  by the trust center, using unsecured key transport when standard security is used or using
  the trust center link key in high-security mode.
    - Standard network keys are used in the standard security mode, and can be used to
      secure general application layer commands.
    - High-security network keys are used in the high-security mode, they are not used for
      communication between secure devices, which use link-specific keys instead.
- **Link keys** can either be negotiated using SKKE, or configured by the trust center under
  request of a device. Service specific keys are derived by hashing of the link key:
    - The **key-load** key is used to protect transported master and link keys.
    - The **key-transport** key is used to protect transported network keys.

A ZigBee device stores a (master key/link key) key pair for each device with which
they may use link-key-based communication. The device is identified by its 64-bit IEEE
address.

7.7.3   The Trust Center

Key distribution is not addressed by 802.15.4. For security purposes, ZigBee defines the
role of "trust center", which is responsible for key distribution and joining policy.

In *high-security mode*, the trust center maintains a list of devices, master keys, link
keys and network keys. A device can be preloaded with the trust center address and
initial master key, or the master key can be sent via an unsecured key transport primitive.
The trust center, by default, is the ZigBee coordinator, but the coordinator can designate
another device, or the trust center can be preconfigured in devices.

In *low-security mode*, a device communicates with its trust center using the current
network key, which can be preconfigured (as in the commercial building automation
profile) or configured via an unsecured key transport primitive during the joining process
(as in the home automation profile).

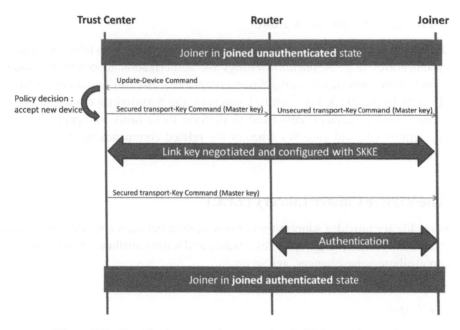

Figure 7.7   Security bootstrapping procedure in high-security mode.

The trust center is notified by ZigBee routers of joining devices by means of an Update-Device command (see Figure 7.7). They can reject or accept the new joiner. In the latter case the trust center communicates the joiner master key to the router, which relays it to the joiner. This enables the trust center to establish a link key to the joiner, and to finally securely communicate the NWK key to the joiner using a secure Transport-Key command (key-transport key, derived from the link key).

Devices can also request the trust center to compute a link key pair for communication with another device: the trust center will communicate the link key to each device using Transport-Key commands.

7.7.3.1   SKKE

Secure ZigBee devices are configured with a link specific master key for each device they may need to communicate securely with, or need to authenticate. A link key, distinct from the master key, can be negotiated between ZigBee endpoints by using a symmetric key establishment (SKKE) scheme using procedures defined in ANSI X9.63-2001.

7.7.3.2   Entity Authentication

ZigBee provides a APSME-AUTHENTICATE.request primitive is used for initiating entity authentication or responding to an entity authentication initiated by another device. The request includes a random challenge and the response is a hash based on the shared master key for the device pair and frame sequence numbers.

7.7.4   The ZDO Permissions Table

The ZDO optionally comprises a permissions configuration table. This table lists a number of tasks categories (e.g., *ApplicationSettings* for authorization to configure bindings, groups, and other application configuration commands) that can be requested from an external entity to the ZDO. For each task category, the permissions configuration table lists the addresses of devices authorized to perform these tasks (or specifies that any device is authorized), and specifies whether the related commands need to be secured with a link-specific key.

7.8   The ZigBee Cluster Library (ZCL)

The cluster library provides support for communication between applications, including over the air configuration of group tables, reading and setting attribute values, subscribing to certain attribute-value changes, and so on.

7.8.1   Cluster

A cluster is a set of related commands and attributes. Each cluster is defined by a ZigBee application profile and identified by a cluster ID (0x0000 to 0xFFFF). Commands are identified by a command ID (0x00 to 0xFF), and attributes by an attribute ID (0x0000 to 0xFFFF).

Clusters are composed of a server side and a client side. Each application residing on a ZigBee device lists the clusters that it supports as a server in its simple descriptor input cluster list, and the clusters that it supports as a client in the simple descriptor output cluster list.

An application that implements the server side typically exposes a set of attributes, and must be able to receive and process the primitives defined by the cluster to read or manipulate these attributes. It may also support attribute-reporting commands, which are sent to requesting devices that implement the client side of the cluster.

Vice versa an application that implements the client side of the cluster may use any of the commands supported by the server side, and must support receiving attribute reporting commands, if it requests such notifications.

Each cluster is defined only within a given application profile. However, in order to avoid any confusion and to redefine common clusters multiple times, the clusters defined by the ZigBee cluster library (ZCL) use the same cluster IDs for each ZigBee public application profile. Table 7.6 lists some of the clusters defined by the ZCL, which therefore have the same meaning for all public application profiles defined by ZigBee.

Clusters are directional: for each endpoint, the endpoint simple descriptor contains the list of input clusters (commands that it accepts and properties that can be written to the endpoint), and the list of output clusters (commands that it may send and properties that may be read).

Table 7.6 Some clusters defined by the ZCL

Cluster ID	Cluster Name
0x0000	Basic cluster
0x0001	Power configuration cluster
0x0002	Temperature configuration cluster
0x0003	Identify cluster
0x0004	Groups cluster
0x0005	Scenes cluster
0x0006	OnOff cluster
0x0007	OnOff configuration cluster
0x0008	Level control cluster
0x000a	Time cluster
0x000b	Location cluster

Each ZigBee public application profile defines a list of devices, each implementing a standard set of features defined by a list of input and output clusters. Within a given public-application profile, manufacturers can extend that list by using the "manufacturer specific extension" field of the ZCL frame (see Section 7.8.4). An endpoint declares the clusters it supports in its simple descriptor, which contains the endpoint number, application profile ID, application device ID, and application version ID (see Section 7.9), the list of input and output clusters.

7.8.2 Attributes

Attributes are identified by a 16-bit number. The ZCL library provides commands to:

- read or set attributes;
- require reporting on an attribute, either periodic or based on a change of value.

7.8.3 Commands

Commands are identified by an 8-bit number. The first 4 bits of cluster-specific commands (defined only within the scope of a given cluster) are set to zero. For instance, command 0x00 of the OnOff cluster (0x0006) means "Off".

Other identifiers are reserved for cross-cluster commands.

7.8.4 ZCL Frame

The ZigBee cluster library frame (Table 7.7) carries application-specific commands.

The general commands (frame type = 00) are listed in Table 7.8.

Table 7.7   ZCL frame format

Field	Bytes	Field details
Frame Control	1	------XX : Frame type (00 : cluster for entire app. Profile, 01 : cluster specific command, other values reserved)
		-----0-- : Manufacturer specific (0 : defined by ZCL, 1 manufacturer specific)
		----X--- : Direction (0 : client to server)
		000X---- : Default response (0 enabled, 1 disabled)
		000----- : Reserved
Manufacturer code	0/2	Present only if the frame is manufacturer specific (Frame control bit 6)
Transaction sequence number	1	Matching of request command frames and response command frames
Command identifier	1	e.g., 0x42: toggle

Table 7.8   ZCL general commands

Command identifier	Command	Usage
0x00	Read attributes	Read a list of attributes, whose 16-bit identifiers are passed as payload
0x01	Read attributes response	Includes a list of attribute records as payload. Each attribute record is as follows: Attribute identifier (16 bit) | Status(Success or error: 8 bit) | Attribute data type (0/8 bit) | Attribute value (variable). Array types are encoded as element type | number of elements | list of elements.
0x02	Write attributes	Write a list of attributes, as described by a list of attribute records in payload.
0x03	Write attributes undivided	Write attributes, all or nothing mode.
0x04	Write attributes response	Response with a list of status codes, for each parameter.
0x05	Write attributes no response	Write attributes, best effort mode.
0x06	Configure reporting	Only specific attributes support reporting in a cluster. The list of attribute reporting configuration records passed as parameter includes minimal and maximal reporting intervals, the minimum change of the attribute that should trigger reporting. This command may be sent by a client to a server to configure reporting, or by a server to a client to describe its future reporting parameters.

(Continued)

Table 7.8 *(Continued)*

Command identifier	Command	Usage
0x07	Configure reporting response	
0x08	Read reporting configuration	Request to get the reporting configuration parameters for one or more attributes
0x09	Read reporting configuration response	
0x0a	Report attributes	Reporting for one or more attributes of a cluster, according to a previously configured reporting relationship (binding)
0x0b	Default response	Basic success/error response
0x0c	Discover attributes	Specifies a starting 16-bit attribute identifier and maximum number of identifiers to report
0x0d	Discover attributes response	
0x0e	Read attributes structured	Used to read only specific index positions (up to 15) for array-type attributes
0x0f	Write attributes structured	Used to write only specific index positions (up to 15) for array-type attributes
0x10	Write attributes structured response	

7.9   ZigBee Application Profiles

An application profile defines a set of messages and attributes standardized for use in a particular context. Application profiles are identified by an application profile ID (0x0000 to 0xFFFF). Each manufacturer can define proprietary messages within its own private application profile, but the ZigBee alliance defines a set of public application profiles, which enable cross-vendor interoperability for the target applications. Profile IDs 0xBF00 to 0xFFFF are reserved for manufacturer specific profiles (MSP), and must be requested from the ZigBee alliance. Profile IDs 0x0000 to 0x7FFF are reserved for public application profiles, and are listed in Table 7.9.

Each ZigBee public profile contains a list of devices identified by a device ID. Each device implements a standard set of features defined by a list of input and output clusters, and a list of attributes (some optional, some mandatory). Within a given public application profile, manufacturers can extend that list by using the "manufacturer-specific extension" field of the ZCL (see Section 7.8.4).

7.9.1   The Home Automation (HA) Application Profile

The HA profile is by far the most widely implemented by manufacturers. While ZigBee does not specify use of nonvolatile memory, the HA profile does: nodes should retain

Table 7.9 ZigBee public application profiles

Public Application Profile		Profile ID	Usage
Home Automation	HA	0x0104	Security, HVAC, LIGHTING CONTROL, ACCESS CONTROL, IRRIGATION...
Commercial Building Automation	CBA	0x0105	Security, HVAC, AMR, lighting control, access control
Industrial Plant Monitoring	IPM	0x0101	Asset management, process control, environmental control, energy management
Telecommunications Applications	TA	0x0107	Information delivery in hot zones, public information enquiry, location-based services, remote control (TV, DVD), cell phone
Automatic (Advanced) Metering Initiative or Smart Energy 1	AMI ZSE 1	0x0109	
Personal Home and Hospital (Health) Care	PHHC	0x0108	Patient monitoring, Fitness monitoring.

their configuration (PAN ID, address, group IDs, etc. ...) even after a power down. This facilitates battery replacement and network restarts after power outages (silent rejoin, see Section 7.3.3).

The HA profile determines the following settings:

- It uses stack profile 0x01 or 0x02.
- Channels 11, 14, 15, 19, 20, 24, 25 are preferred.
- The trust center link key is hardcoded in the profile, the trust center distributes a network key.
- The minimum number of entries of entries in the broadcast transaction table (BTT, see Section 7.4.5.1) is nine.

The standard devices defined by the home automation public application profile are characterized by the mandatory clusters that they must support. Mandatory and optional "common clusters" are defined for all HA devices:

- Server side: mandatory support of basic and identify clusters. Optional support for power configuration, device temperature configuration, alarms, meter, and manufacturer-specific clusters.
- Client side: optional support for meter, and manufacturer-specific clusters.

The HA profile also lists mandatory and optional clusters specific to each device type. Table 7.10 lists some examples.

Table 7.10 Some devices defined by the HA public application profile

	Device name	Device ID	Supported clusters Mc/Ms: mandatory on client or server side, Oc/Os: optional on client or server side
Generic	On/Off switch	0x0103	Ms: OnOff (0x0006) Os: OnOffSwitch Config (0x0007) Oc: Scenes (0x0005) Oc: Groups (0x0004) Oc: Identify (0x0003)
	Range Extender	0x0008	Only common clusters
	Mains Power Outlet	0x0009	Ms: OnOff (0x0006) Ms: Scenes (0x0005) Ms: Groups (0x0004)
Lighting	On/Off Light	0x0100	Ms: OnOff (0x0006) Ms: Scenes (0x0005) Ms: Groups (0x0004)
	DimmableLight	0x0101	Ms: OnOff (0x0006) Ms LevelControl (0x0008) Ms: Scenes (0x0005) Ms: Groups (0x0004) Oc: Occupancy sensing (0x0406)
	Light Sensor	0x0106	Ms: Illuminance measurement (0x0400) Oc: Groups (0x0004)
	DimmerSwitch	0x0104	Mc: OnOff (0x0006) Mc LevelControl (0x0008) Oc: On Off switch configuration (0x0007) Os: Scenes (0x0005) Os: Groups (0x0004)
Closures	Shade	0x0200	Ms: OnOff (0x0006) Ms LevelControl (0x0008) Ms: On Off switch configuration (0x0007) Ms: Scenes (0x0005) Ms: Groups (0x0004)
	Shade Controller	0x0201	Mc:OnOff (0x0006) Mc LevelControl (0x0008) Oc: Shade configuration (0x0100) Oc: Scenes (0x0005) Oc: Groups (0x0004) Oc: Identify (0x0003)

Table 7.10 (*Continued*)

	Device name	Device ID	Supported clusters Mc/Ms: mandatory on client or server side, Oc/Os: optional on client or server side
HVAC	Heating / Cooling unit	0x0300	Ms:OnOff (0x0006) Mc: Thermostat (0x0201) Os: Fan control (0x0202) Os: Level control (0x0008) Os: Groups (0x0004)
	Thermostat	0x0301	Ms: Thermostat (0x0201) Os: Scenes (0x0005) Os: groups (0x0004) Os: Thermostat user interface configuration (0x0204) Os/ Oc: Fan control (0x0202) Os / Oc: Temperature measurement (0x0402) Os / Oc:: Occupancy sensing (0x0406) Os/Oc: Relative humidity measurement (0x0405)
	Temperature sensor	0x0302	Ms: Temperature measurement (0x0402)

7.9.2   ZigBee Smart Energy 1.0 (ZSE or AMI)

ZigBee Smart Energy 1.0 is a public application profile (profile 0x0109), documented in "ZigBee SMART ENERGY PROFILE SPECIFICATION (rev 15, 1/12/2008)". It defines the smart energy devices and clusters required to build an energy-management system (EMS).

The ZigBee Smart Energy 1 (ZSE) public application was defined to enable usage of ZigBee for automatic metering, demand response and prepayment applications required by utilities. The ZSE was defined just before ZigBee decided that its next version would rely on IP, and ZSE was the first application profile to be entirely redesigned in the context of IP, the new specification is called ZigBee Smart Energy 2.0 (see Section 13.4).

ZSE dedicates a secure HAN (use of security is mandatory) to the utility, and defines the communication primitives used between the energy service portal controlled by the utility, and devices located in the end user primitives, such as home displays or load control devices. As the displays and other ZigBee devices may be deployed by end users on their home network using HA, ZSE also defines an extension (the "stub APS") to the ZigBee core specification enabling limited single hop communications between two PANs (the utility PAN and the user home area network).

ZSE defines several new clusters listed in Table 7.11.

7.9.2.1 Security

SE makes a mandatory use of link keys (preconfigured or commissioned), which are otherwise optional in ZigBee. However, a master key is not used or preconfigured in ZigBee SE devices, which do not operate in "high-security" mode.

ZigBee smart energy devices use a network key allocated by the trust center, using the key establishment cluster with a preconfigured trust center link key. The link key is also replaced by the trust center as part of this security bootstrapping process. ZigBee smart energy networks will not generally send keys in the clear.

ZigBee SE envisions that two separate home area networks will be used:

- One network for the exclusive use of the utility company, interconnecting the energy services portal, in-home display(s) and load control devices.
- One separate network for the home owner use, including home automation devices, in-home displays (ihd), a home energy management console, and smart appliances.

The links between the automatic metering infrastructure (AMI) servers and the home area networks may use a combination of non-ZigBee and ZigBee networks. The energy services portal may control a collection of sub-ESPs, in a cascading fashion, so that ZigBee can optionally also be used as a neighborhood area network (NAN).

Since all ZigBee SE devices are configured with multiple keys (link keys and the network key), ZigBee SE defines which cluster uses which key. All the ZigBee SE clusters use application link keys, but most general clusters (e.g., basic cluster, alarm cluster, identify cluster) use the network key.

7.9.2.2 Smart Energy Extensions of ZigBee

ZigBee SE defines a simplified APS layer designed for basic "inter-PAN" communication, that is, communication between a PAN and a device that has not joined. The specification mentions the "refrigerator magnet" with an LCD screen as a target.

Such messages can be sent unicast, broadcast or sent to a group, but without any security.

7.9.2.3 Smart Energy Devices

ZigBee SE defines the following "smart energy" devices:

Energy Service Portal (ESP, Device Id 0x0500)
The energy service portal is a server controlled by the utility company that connects to the metering and energy management devices within the home. The ESP acts as the coordinator and trust center of the network.

The ESP must support the server side of the price, message, demand response/load control and time clusters, and optionally of the complex metering, simple metering and prepayment. It may support the client side of the simple metering, complex metering, price and prepayment clusters.

Metering Device (Device Id 0x0501)
A metering device must support the client side of the metering cluster, optionally of the complex metering cluster. It may support the client side of the metering prepayment, price and message clusters.

In-Premises Display Service (Device Id 0x0502)
The device is designed to be used as a simple user interface, displaying graphs or messages, and able to signal button press events. It must implement the client side of at least one of the price, simple metering and messaging clusters.

Programmable Communicating Thermostat (PCT, Device Id 0x0503)
This device is designed to control heating and cooling systems. It must implement the client side of the Demand response/load control and time clusters. It may implement the client side of the prepayment, price, simple metering and message clusters.

Load Control Device (Device Id 0x0504)
This device is designed to manage the electric consumption of devices in a generic way. It must implement the client side of the demand response/load control and time clusters. It may implement the client side of the price message cluster.

Range Extender Device (Device Id 0x0008)
A device announcing this device Id must be a pure ZigBee router, not supporting any other function.

Smart Appliance Device (Device Id 0x0505)
Smart appliances are able to participate in energy-management policies. They must implement the client side of the price and time clusters, and optionally of the demand response/load control and message clusters.

Prepayment Terminal Device (Device Id 0x0506)
This device is not fully specified yet. It is designed to support advance payment of services, and various display functions.

7.9.2.4   Smart Energy Clusters

Demand Response and Load Control Cluster (0x0701)
Server-Side Commands

Table 7.11   New clusters defined by ZigBee smart energy

Price	0x0700
Demand response and load control	0x0701
Simple metering	0x0702
Message	0x0703
Registration	0x0704
AMI tunneling (complex metering)	0x0705
Prepayment	0x0706

Load control event (0x00). This event is sent to the devices asked to implement a load control action, and specifies the actions required. It includes the following parameters:

- *Issuer Event ID*: unique identifier that will be used to identify future event reports related to this demand response and load control event.
- *Device Class*: bit-encoded field indicating the device classes (end devices actually performing the energy-demand response, as opposed to their ZigBee controllers) needing to participate in an event. The classes defined include HVAC compressors, Strip heaters, water heaters, electric vehicles, and so on.
- *Utility Enrolment Group*: both the utility enrolment group field and the device class must match the target device configured values, otherwise it will ignore the command.
- *Start Time*: UTC Timestamp or 0x00000000 for "now."
- *Criticality Level*: levels 1 to 9 are currently defined. Participation in levels 1 to 6 is voluntary, participation in level 9 is mandatory. Level 1 signals an abnormal percentage of nongreen sources in the delivered energy.
- *Cooling and heating temperature offsets*: offsets, in units of 0.1 °C, to the local temperature setpoint of the thermostat (added for cooling systems, subtracted for heating systems), noncumulative across sequential demand response events. 0xFF when not used.
- *Cooling and heating temperature setpoint*: request to replace the current setpoint by the indicated value between −273 °C and 327 °C in units of 0.01 °C, set to 0x8000 when not used.
- *Average Load Adjustment Percentage*: defines a load offset of −100 to +100 points relative to 100%, with a resolution of 1 point. Interpretation is specific to the client implementation. Set to 0x80 when not used.
- *Duty cycle*: a percentage of time between 0 and 100%, 0xFF when not used.
- *Event Control*: flags to indicate whether randomization of the start and end times is required.

Cancel load control event (0x01): cancellation order specifying the issuer event Id, utility enrollment group and device classes concerned. Flag to optionally override the end randomization settings of the original load control event, and desired cancellation UTC time (0x00000000 for "now").

Cancel all load control events (0x02): same as the cancel load control event command, without the filtering parameters.

Client Side

The client side of the load control/demand response cluster must be able to store at least 3 scheduled events. Events exceeding the storage capacity should be retrieved as soon as possible using command "get scheduled events".

The client side maintains several attributes:

- The utility enrolment group;
- The start randomization period, and the stop randomization period, in minutes;
- The device class bitmask.

It implements two commands:

- **Get Scheduled Events (0x01),** which asks the server side to resend up to a certain number of load control commands scheduled to start at or after the specified UTC time stamp.
- **Report Event Status (0x00),** which reports the time at which the load-control event (identified by its event ID) was effectively executed, the criticality level, the cooling or heating set points, load adjustment percentage, duty cycle that were applied. The event status parameter contains the current state of the load control event: started, completed, user opted in or opted out before or during the event, canceled, superseded, and error conditions. An electronic signature ensures nonrepudiation.

Simple Metering Cluster (0x0702)
Server-Side Attributes

Attribute set identifier	Attributes	
0x00 Reading information set	CurrentSummationDelivered (0x00), CurrentSummationReceived (0x01),	Most recent summed value of energy/gas/water delivered to, or provided by the premises.
	CurrentMaxDemandDelivered (0x02), CurrentMaxDemandReceived (0x03), CurrentMaxDemandDeliveredTime (0x08), CurrentMaxDemandReceivedTime (0x09)	Instant value of maximum rates, and when they were measured.
	DFTSummation (0x04), DailyFreezeTime (0x05),	Snapshot of CurrentSummationDelivered taken at instant DailyFreezeTime

	PowerFactor (0x06), ReadingSnapShotTime (0x07),	Average power factor
0x01 TOU information Set	CurrentTierNSummationReceived CurrentTierNSummationDelivered	Where $N = 1$ to 6, partial summation counters per price tier (defined per period in the time of use schedule or per price tier).
0x02 Meter status	Status (0x00)	Status flags: tamper detected, leaks, etc.
0x03 Formatting	UnitOfMeasure (0x00), Multiplier (0x01), Divisor (0x02), SummationFormatting (0x03), DemandFormatting (0x04), HistoricalConsumptionFormatting (0x05),	Set of attributes used to transform counters to displayable values: units, scaling factors.
0x04 ESP historical consumption	InstantaneousDemand (0x00), CurrentDayConsumptionDelivered (0x01),CurrentDayConsumptionReceived (0x02), PreviousDayConsumptionDelivered (0x03), PreviousDayConsumptionReceived (0x04), CurrentPartialProfileIntervalStartTimeDe-livered (0x05), CurrentPartialProfileIn tervalStartTimeReceived (0x06), CurrentPartialProfileIntervalValueDeliv-ered (0x07), CurrentPartialProfileIntervalValueReceived (0x08)	Accumulators since midnight for the current day, since the start time of the current profile interval, for the previous day
0x05 Load profile configuration	MaxNumberOfPeriodsDelivered (0x00)	maximum number of intervals the device is capable of returning in one get profile response command.
0x06 Supply Limit	CurrentDemandDelivered (0x00), DemandLimit (0x01), DemandIntegrationPeriod (0x02), NumberOfDemandSubintervals (0x03)	Demand is integrated during the demand integration period, and the integration result is written to currentDemandDelivered at the end of each subinterval.

Server-Side Commands

GetProfileResponse (0x00) returns a number of period accumulators for periods ending before a specified EndTime.

    RequestMirror (0x01): request the ESP to mirror metering device data, using RequestMirrorResponse command.

    command

    RemoveMirror (0x02): request the ESP to remove its mirror of metering device data.

Client-Side Commands

GetProfile (0x00): requests a number of period accumulators for periods ending before a specified EndTime, for received or for delivered quantities.

    RequestMirrorResponse (0x01): the ESP informs a sleepy metering device it has the ability to store and mirror its data, which should be sent to the indicated endpointID.

    RemoveMirror (0x02): the ESP no longer has the ability to mirror data.

Price Cluster (0x0700)

Server-Side Attributes

6 price tiers labels are defined, by attributes "TierXpriceLabel" (0x0000 to 0x0006). The price tiers are defined by command publish price.

Client-Side Commands

- **getCurrentPrice (0x00)**: initiates a publish price command for the current time.
- **getScheduledPrices (0x01)**: initiates a publish price command for all currently scheduled times after the provided time stamp (up to a maximum number of scheduled times also specified in the command).

Server-Side Commands

Publish price (0x00): this command defines a new price tier and is sent in response to a getCurrentPrice or getScheduledPrices command. It contains the following subfields:

- *ProviderID:* unique Id of the commodity provider;
- *Rate Label:* 12 character UTF8 string related to current rate;
- *Issuer Event ID:* unique identifier for this pricing information, must increase when newer prices are published for the same period;
- *Current Time:* UTC time of the sending node;
- *Unit of measure:* 8-bit field defining the commodity and unit of measure;
- *Currency*
- *Price Trailing Digit and Price tier:* bit field indicating the price tier for this rate, and the position of the decimal point in the published price;

- *Number of price tiers and Register tier:* number of price tiers in use (0 to 6), for the current tier, indication of which CurrenttierXSummationDelivered accumulator relates to the current price tier;
- *StartTime:* UTC start time of the current rate signal, 0x00000000 means "now";
- *Duration in minutes:* validity of this price signal. 0xffffffff means "until changed";
- *Price:* price per base unit;
- *PriceRatio* (optional): ratio to the "normal" tariff;
- *Generation Price* (optional): price for commodity received from the premises;
- *AlternateCostDelivered*, Alternate cost unit and alternate cost trailing digit: cost using an alternate measure, for example, grams of CO_2 per kWh.

Messaging Cluster (0x0703)
Server-Side Commands
DisplayMessage (0x00): specifies the level of importance of the message, whether a confirmation is required, and the number of minutes the message must be displayed.
   CancelMessage (0x01)

Client-Side Commands
GetLastMessage (0x00): request to send a DisplayMessage command
   MessageConfirmation (0x01): used to acknowledge a message, provides a timestamp.

Smart Energy Tunneling (Complex Metering) Cluster (0x0704)
Not defined yet, this cluster is a placeholder for future tunneling mechanisms for more sophisticated metering protocols, for example, C.12 or DLMS/COSEM.

Prepayment Cluster (0x0705)
Not defined yet.

7.10   The ZigBee Gateway Specification for Network Devices

The ZigBee Gateway specification for network devices, Version 1.0, was released on July 27th 2011. It had been a work in progress since 2007. This specification defines several possible communication protocols between ZigBee gateway devices (ZGD) that implement a bidirectional interface between ZigBee 1.0 PANs and IP networks, and IP host applications (IPHA).
   The ZGD provides access to:

- ZCL operations: read and write attributes, configure notifications and report events;
- ZDO operations;
- Macro operations simplifying network and service discovery;
- Endpoint management;

- Its own ZigBee information bases (AIB, NIB, and PIB attributes);
- Network startup and join functions on behalf of the IPHA;
- Security material configuration and operations.

The communication between ZGDs and IPHAs is bidirectional, both act as client and server. The IPHA calls procedures implemented by the ZGD, and the ZGD calls event handlers of the IPHA. These RPC functions have been specified in an abstract, protocol-independent request-response format that needs to be complemented by a protocol binding specification.

Some ZGD procedures operate only in blocking mode, while other procedures offer a choice of blocking or nonblocking mode (such nonblocking procedures provide a *CallbackDestination*, which is an IPHA callback URL or an empty string if the IPHA uses polling). All IPHA event handlers are nonblocking.

ZigBee PAN messages received by the ZGD are processed by one or more ZGD callback handlers, which decode and feed them to requesting IPHAs. The configuration of these callback handlers is persistent across ZGD power cycles.

Version 1.0 of the ZigBee gateway specification proposes three possible network bindings implementing the ZGD to IPHA remote procedure calls: a SOAP binding, a REST binding, and a GRIP (gateway remote interface protocol) binding.

7.10.1   The ZGD

The ZGD itself is a ZigBee device. As such it should support the ZigBee commissioning cluster, as well as mandatory ZDO client and server clusters.

The ZGD functions belong to several categories: gateway management object, Zig-Bee device profile, ZigBee cluster library, application support sublayer, inter-PAN, and network layer.

An IPHA can access:

- APS commands to read and modify the ZGD configuration by using a set of APS functions. It can read or set descriptors (ConfigureNodeDescriptor, GetNodeDescriptor, GetNodePowerDescriptor, ConfigureUserDescriptor, GetUserDescriptor), manage local ZGD endpoints (ConfigureEndpoint, ClearEndpoint), send and receive APS frames (SendAPSMessage / NotifyAPSMessageEvent), manipulate local groups (AddGroup, RemoveGroup, RemoveAllGroups, GetGroupList), and read local bindings (GetBindingList).
- Network layer commands to get access to the Network layer management entity (NLME) of the ZGD (FormNetworkProcedure, FormNetworkEvent, StartRouter, StartRouterEvent, Join, JoinEvent, Leave, LeaveEvent, Reset, ResetEvent, Discover-Networks, DiscoverNetworksEvent, PerformEnergyScan, PerformEnergyScanEvent,

NetworkStatusEvent, PerformRouteDiscovery, PerformRouteDiscoveryEvent, Send-NWKMessage, NotifyNWKMessageEvent).

An IPHA can send arbitrary ZDP frames by using the SendZDPCommand ZGD function and receive arbitrary ZDP frames by implementing a NotifyZDPEvent event handler.

The ZGD also provides access to the ZCL. An IPHA can send and receive arbitrary ZCL commands by using the SendZCLCommand ZGD function and implementing the NotifyZCLEvent event handler.

In addition the ZGD supports the specific gateway management object (GMO). The GMO provides functions:

- Enabling the IPHA to retrieve the ZGD version and feature set (GetVersion).
- Enabling read/write of ZGD information base (Get/Set).
- Event CallBack management (CreateCallBack, DeleteCallBack, ListCallBacks) and polling by IPHA (PollCallBack, PollResults, UpdateTimeOut).
- Facilitating network and service discovery (StartNodeDiscovery, NodeDiscoveryEvent, NodeLeaveEvent, ReadNodeCache, StartServiceDiscovery, ServiceDiscoveryEvent, GetServiceDescriptor, ServiceDiscoveryEvent, ReadServiceCache).
- Controlling the Gateway insertion into the PAN (StartGatewayDevice, StartGateway-DeviceEvent, ConfigureStartupAttributeSet, ReadStartupAttributeSet).
- Managing address aliases (CreateAliasAddress, DeleteAliasAddress, ListAddresses): In order to facilitate the mapping of addresses between the PAN network (short 16-bit addresses may change in the PAN) and the IPHA, the ZGD maintains an address alias table where IPHAs can associate their own alias to any 64-bit address. The IPHA or ZGD may use special code 0x10 in the Destination/source address mode[2] of ZDP messages to indicate use of an alias address: the ZGD will translate to 64-bit or 16-bit addresses as required.

7.10.2   GRIP Binding

The gateway remote interface protocol (GRIP) is a lightweight request/response protocol built over SCoP.

> The secured connection protocol (SCoP) was designed in the context of the ZigBee bridge device specification as an IP tunneling protocol between ZigBee PAN gateways. It provides support for datagram and stream-oriented communications as well as fragmentation by means of a socket-like

---

[2] **APSDE-DATA.request DstAddrMode parameter values :** 0x00 = DstAddress and DstEndpoint not Present; 0x01 = 16-bit group address for DstAddress; DstEndpoint not present; 0x02 = 16-bit address for DstAddress and DstEndpoint present; 0x03 = 64-bit extended address for DstAddress and DstEndpoint present; 0x04 – 0xff = reserved.

interface provided by the SCoP data entity (SCDE), on top of UDP, TCP or TLS. SCoP security leverages CCM* (SCoP security service SCSS). Management services are provided by the SCoP management entity (SCME).

SCoP messages target a specific service (identified by a service identifier code), on the target gateway. Currently the supported services are SCoP service (hello, goodbye, keepalive commands), bridge service and GRIP service.

The GRIP API provides a GRIDE-DATA request that must be provided with target information (DestIPVersion, DestIPAddress, DestPort), desired transport parameters (TransportMode, SecurityLevel) and application-level parameters (target FunctionDomain, FunctionCategory, ManufacturerCode, FunctionIdentifier, Function-Parameters).

Requests are provided to the target GRIP application by means of the GRIDE-DATA.indication API. Responses are returned to the querying GRIP application by means of the GRIDE-DATA.response API, and specifies a Status code as well as a function result (an octet string).

The GRIP binding specifies specific GRIP function codes and parameter formats to transport all the generic ZGD procedures defined in Section 7.10.1. The following function categories are defined: Manufacturer specific, GMO, ZDP, ZCL, APS, INTERPAN, NWK. Parameter encoding uses ASN.1 distinguished encoding rules (DER).[3]

7.10.3 SOAP Binding

The SOAP binding specifies a standard web services interface by means of a WSDL document. All the generic ZGD procedures defined in Section 7.10.1 are covered.

7.10.4 REST Binding

Due to the resource oriented design pattern of REST, the REST binding for the ZDG functional protocol is not as straightforward as the GRIP or SOAP bindings.

A set of resources are defined to represent the ZGD (Figure 7.8), each ZigBee network and each ZigBee node (Figures 7.9 and 7.10), each resource is addressable with an

---

[3] ISO 8825, 1998: Information Technology. ASN.1 Encoding rules: Specification of Basic Encoding Rules (BER), Canonical Encoding Rules (CER) and Distinguished Encoding Rules (DER). International Standard ITU-T X690 (1997) | ISO/IEC 8825-1:1998.

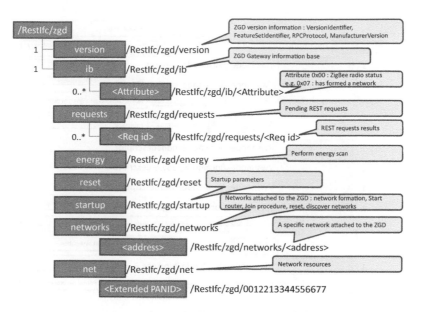

Figure 7.8 ZGD REST resources overview.

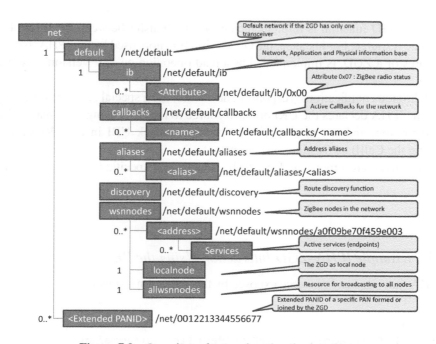

Figure 7.9 Overview of network and node resources.

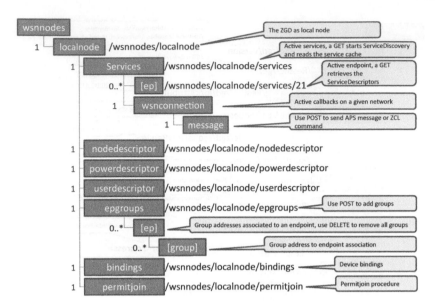

Figure 7.10 Overview of ZigBee node resources.

URL. Each ZGD function is implemented as a state transfer between the IPHA and the appropriate resource: For instance, the GetVersion function, used by the IPHA to discover the features of a ZGD, is implemented as a read operation on the "version" resource of the ZGD.

Asynchronous notifications (responses and events) use a special resource hosted by the IPHA. The IPHA declares the URI of this resource to the ZGD using the CreateCallBack REST operation, and can also specify a specific callback URI in each REST request supporting the CallBackDestination parameter.

The resources are represented by XML documents, specified as part of the REST binding.

7.10.5 Example IPHA–ZGD Interaction Using the REST Binding

The IPHA may read any resource exposed by the ZGD, for instance

```
HTTP GET:
http://''ZGD_IP_Addr:ZGD_PORT''/RestIfc/zgd/version
```

And the ZGD responds with the requested resource representation

```
<tns:Info xmlns:tns=''http://www. zigbee.org/GWGSchema'' >
    <tns:Status>
        <tns :code>0x00</tns :code>
    </tns:Status>
    <tns:Detail>
        <tns:Version>
            <tns:VersionIdentifier>0x01</tns:VersionIdentifier>
            <tns:FeatureSetIdentifier>0x00</tns:
                FeatureSetIdentifier>
            <tns:RPCProtocol>0x0004</tns:RPCProtocol>
            <tns:ManufacturerVersion>1.1</tns:ManufacturerVersion>
            </tns:Version>
        <tns :Detail>
</tns:Info>
```

In this example we suppose that the IPHA needs to allocate a dedicated endpoint on the ZGD. It therefore begins by registering a new endpoint on the ZGD. In the following example wc usc a synchronous transaction. The IPHA could have use an asynchronous method by specifying a callback URI in parameter *CallbackDestination* of request URI sent to the ZGD.

```
HTTP POST: HTTP POST: http://''ZGD_IP_Addr:ZGD_PORT''/
localnode/services
```

```
<?xml version=''1.0'' encoding=''UTF-8''?>
<tns:SimpleDescriptor
xmlns:tns=''http://www. zigbee.org/GWGSchema'' >
<tns:ApplicationProfileIdentifier>0x0104</tns:ApplicationProfile
Identifier>
<tns:ApplicationDeviceIdentifier>0x0002</tns:ApplicationDevice
Identifier>
<tns:ApplicationDeviceVersion>0x00</tns:ApplicationDevice
Version>
<tns:ApplicationInputCluster>0x0000</tns:ApplicationInput
Cluster>
<tns:ApplicationInputCluster>0x0003</tns:ApplicationInput
Cluster>
<tns:ApplicationInputCluster>0x0004</tns:ApplicationInput
Cluster>
<tns:ApplicationInputCluster>0x0005</tns:ApplicationInput
Cluster>
<tns:ApplicationInputCluster>0x0006</tns:ApplicationInput
Cluster>
<tns:ApplicationOutputCluster>0x0003</tns:ApplicationOutput
Cluster>
</tns:SimpleDescriptor>
```

The IPHA did not request a specific endpoint, therefore it will be allocated by the ZGD. If the request succeeds, the IPHA will receive a response similar to the following:

```
<tns:Info xmlns:tns=''http://www. zigbee.org/GWGSchema'' >
    <tns:Status>
        <tns :code>0x00</tns :code>
    </tns:Status>
    <tns:Detail>
        <tns:endpoint>0x23</tns:endpoint>
    <tns :Detail>
</tns:Info>
```

This is an example of a synchronous response. When using an asynchronous mode, the response would include a request identifier attribute allocated by the ZGD.

The IPHA has the option of registering a callback function bound to this endpoint or to all endpoints, by invoking an HTTP POST request to [NwkRootURI]/callbacks that specifies filters which will be used by the ZGD to identify messages that will be notified to the IPHA.[4]

The IPHA may then instruct the ZGD to send any command. The endpoint identifier used in the URI is 35 (0x23). For instance an APSMessage:

```
POST [NwkRootURI]/localnode/services/35/wsnconnection/message
<?xml version=''1.0'' encoding=''UTF-8''?>
<tns:APSMessage xmlns:tns=''http://www.zigbee.org/GWGSchema''>
    <tns:DestinationAddress>
        <tns:NetworkAddress>0x0001</tns:NetworkAddress>
    </tns:DestinationAddress>
    <tns:DestinationEndpoint>0x02</tns:DestinationEndpoint>
    <tns:SourceEndpoint>0x23</tns:SourceEndpoint>
    <tns:ProfileID>0x0104</tns:ProfileID>
    <tns:ClusterID>0x0000</tns:ClusterID>
    <tns:Data>01020304050607008090a0b0c0d0e0f</tns:Data>
    <tns:TxOptions>
    <tns:SecurityEnabled>true</tns:SecurityEnabled>
    <tns:UseNetworkKey>true</tns:UseNetworkKey>
    <tns:Acknowledged>true</tns:Acknowledged>
    <tns:PermitFragmentation>true</tns:PermitFragmentation>
    </tns:TxOptions>
    <tns:Radius>3</tns:Radius>
</tns:APSMessage>
```

---

[4] Callbacks may also be registered using resources [NwkRootURI]/localnode/services/[ep]/wsnconnection or [NwkRootURI]/localnode/allservices/wsnconnection, in which case filters are preset for all APS messages matching the endpoint id.

If the APS command is successful, the IPHA will receive the following response from the ZGD:

```
<tns:Info xmlns:tns=''http://www. zigbee.org/GWGSchema'' >
    <tns:Status>
        <tns :code>0x00</tns :code>
    </tns:Status>
    <tns:Detail>
        <tns: APSMessageResult xmlns:tns=''http://www. zigbee.
        org/GWGSchema''>
            <tns:ConfirmStatus>0x00</tns:ConfirmStatus>
            <tns:TxTime>0x01234567</ts:TxTime>
        </tns:APSMessageResult>
    <tns :Detail>
</tns:Info>
```

8

Z-Wave

8.1 History and Management of the Protocol

The Z-wave protocol was designed by private company Zensys (www.zen-sys.com), based in the US and Denmark. Zensys is now a subsidiary of SIGMA Designs, a provider of system on chip (SoC) products for the multimedia and entertainment industry. Zensys started by introducing to the market, in 2001, a light-control system for consumers, and evolved its product to a full-fledged home area network meshed protocol implemented in a proprietary SoC.

Z-wave quickly became a very popular home automation protocol, with hundreds of products sold on web sites like www.zwaveproducts.com or www.zwaveworld.com. Zensys customers appreciate that Z-wave offers approximately the same features as its standard competitor 802.15.4 (ZigBee), but with fewer interoperability issues (due to the multi-vendor nature of the ZigBee ecosystem), and an unlicensed frequency band (868 MHz) that is often perceived to be less problematic than the crowded 2.4 GHz band.

Zensys OEMs using the protocol are grouped in the Z-wave alliance (www.z wavealliance.org), which promotes awareness of the product, organizes developer's forums and interoperability testing events ("Unplugfest"). Z-wave products are certified by Zensys or agreed labs, using tools provided by Zensys, which test various network management functionalities and assess the communication error rate (CER) of the device under test at various distances.

The evolution of the system is managed by Zensys technical service (TS, http://support.zen-sys.com). This section describes the third generation of the chip and protocol stack, as available in March 2010 (ZW0302 SoC). There is another version in preparation, but not yet released.

The Internet of Things: Key Applications and Protocols, First Edition.
Olivier Hersent, David Boswarthick and Omar Elloumi.
© 2012 John Wiley & Sons, Ltd. Published 2012 by John Wiley & Sons, Ltd.

8.2   The Z-Wave Protocol

8.2.1   Overview

In the design of Z-wave, Zensys targeted a precise market segment: low-cost and reliable home area networks. In other words "better that X10"[1] (that means reliable), but still low cost: Zensys targets the 3 USD per SoC price segment, about 30% of the cost of RF systems capable of multimedia communications such as WiFi.

The chip has been optimized for battery-powered nodes: it remains in "power down" mode (consuming about 2.5 μA for the ZW0302 SoC) most of the time, and "Wakes-up" only for brief intervals from time to time to perform its function, receive and transmit. Wake-up occurs at programmable intervals or if an interrupt is raised, and at these moments the SoC power consumption is of course higher (about 25 mA when receiving or transmitting).

The Z-wave offering, as provided by Zensys developer's kits to its customers, covers the physical layer (RF link), the link and routing layers (called by Zensys the MAC, transfer and routing layers). The transport layer is a simple resend scheme, and there is no concept of a session. APIs are provided at the routing-layer level for application developers.

RF home area networks typically use very low transmission power (many RF nodes are battery powered), in a complex environment with multiple reflections of the signal, diffraction, and so on. In this context, the strategy selected by Zensys for Z-wave is to build a mesh network, so that nodes with no power-consumption constraints can relay the radio signals from room to room. In a mesh network, there is no need for a node to be able to communicate directly with any other node: the mesh protocol builds routing tables ensuring that any node has a valid path, via selected nodes, to any other node in the network.

8.2.2   Z-Wave Node Types

From a *radio perspective*, there are two types of nodes:

- Devices that can enter sleep mode. All battery-powered devices are in this situation. They cannot be used by the Z-wave network as relays to other nodes.
- *Always listening* nodes, identified by a specific flag, can be used as repeaters in the Z-wave networks because their radio is always listening. Usually, most devices powered from AC mains are "always listening".

From a *radio-routing perspective*, Z-wave defines two main types of nodes:

- *Controllers*, which have and maintain the full topology of the network, and can assign routes to slaves;

---

[1] X.10 is a popular CPL-based low-cost home automation protocol, also known for its unreliability (commands are not acknowledged and easily affected by parasitic noise on power lines).

– *Slaves*, which have a limited knowedge of the network topology and no functionality related to the maintenance of the network topology (e.g., including or excluding nodes).

8.2.2.1 Controllers

Controllers maintain a complete network topology map, and therefore can calculate routes to reach any node in the network. Zensys defined several subtypes of controllers:

- *Primary Controller*. A controller that has the capability to include or exclude nodes in the network, maintains the list of nodes in the network and calculates the route table.
- *Secondary Controller*. A controller that receives a copy of the list of nodes and of the route table. In the case of a failure of the primary controller, a secondary controller can take the role of a primary controller.

Some other variants have appeared in successive versions to fulfill more evolved needs:

- *Portable Controller*. A controller that is typically mobile and battery powered, therefore not always listening. In the first generations of Z-wave networks, the node inclusion was performed at a reduced radio power, therefore there was a need for a portable controller. In more recent networks, other strategies can be used (inclusion controller, inclusion at full power or explorer frame, see below) and therefore a portable controller is not required.
- *Static Controller*. Installed in a fixed location and mains powered, a static controller is always listening. A static controller can be primary, but should preferably be a secondary controller if the network uses low power inclusion, due to range limitations. A static controller can optionally be configured as:
  - a *Static Update Controller* (SUC): a SUC is a network database that gets updates from the primary controller about all network changes, and can send network topology information to secondary controllers and routing slaves that request updates. The SUC function is assigned to a static controller by the primary controller. Secondary controllers with SUC functionality can take over the role of the primary controller if it fails.
  - *SUC ID Server* (SIS). This optional functionality of a SUC was introduced with Zensys Developer's Kit 3.40. The SIS enables other controllers to include/exclude nodes in the network on its behalf: when a SIS is present all controllers will become *inclusion controllers* instead of primary and secondary controllers. They can include/exclude nodes to a network and provide correct home ID and node ID as long as these inclusion controllers can reach the controller with SIS functionality.
- *Bridge Controller*: designed to control up to 128 devices located on other automation networks (e.g., LonWorks, BACnet, KNX or X.10) by emulating as many virtual slave Z-wave nodes itself.

8.2.2.2 Slaves

A Z-wave node that is mains powered (always listening) and is able to execute or route commands from other Z-wave devices in the network, with no other network functionalities, is a **Slave**. Typically, a Z-wave power plug is a slave node. A slave device must be in a fixed position and must be listening at all times. Slave devices have no topology map and cannot calculate routes. Since they are always listening, slaves can receive commands via routes (from controllers or routing slaves) and send replies back using the same route (the *response route*). Slaves must act as repeaters in the network, forwarding messages to the next hop that must be provided in the route that is part of the message.

A node that can send unsolicited messages to other nodes in the network is a ***routing slave***, for example a battery-powered temperature sensor. Routing slaves can receive commands and send replies, but they have a partial topology map and cannot calculate routes. Therefore, they can only address a subset of the network to send unsolicited frames (up to 5 nodes), and must store static routes to the targets of the unsolicited messages. They can also store several routes to the static controller with SUC/SIS functionality and ask for route updates. Routing slaves can optionally act as a repeater (*slave*) in the network if they are mains powered.

Z-wave defines 3 subtypes of routing slaves:

- *FLiRS Routing Slave*. Frequently listening routing slave (FLiRS). This is a normal routing slave configured to listen for a wakeup beam in every wake-up interval. This enables other nodes to wake up the FLiRS and sends a message to it. On only ZDK 5.0x supports FLiRS nodes. One usage for a FLiRS could be as the chime node in a wireless doorbell system.
- *Enhanced Slave*: is a routing slave, with an external EEPROM to store application data.
- *ZensorNet^{TM} Routing Slave*: these nodes add low latency point to point communication support for battery-powered nodes in a Z-Wave® network, as required by simple smoke-alarm applications. The ZensorNet routing slave node is basically a routing slave node configured as FLiRS with the additional functionalities of Zensor net binding and Zensor net flooding. See Section 8.2.4.1 for details.

8.2.3   RF and MAC Layers

Z-wave chips uses a B-FSK "spectrum shaping" modulation over a 868.42 MHz carrier frequency in the EU, 908.42 MHz in the USA, 921.42 MHz in Australia, and 919.82 MHz in Hong Kong.

The chip output power is programmable from -20 to 0 dBm, and the chip is able to decode signals as low as -102 dBm for a data rate of 9.6 kbps (about 30 m indoors, and 150 m in line of sight conditions). The maximum bitrate attainable with the current version is 40 kbps in favorable link conditions. It is lower than the maximum bitrate of 802.15.4 (ZigBee) systems, which operate around 100 kbps. However, Zensys plans to add support for other frequencies (950 MHz and 2.4 GHz) in Z-wave 6.00, which also targets a higher maximum bitrate over 100 kbit/s.

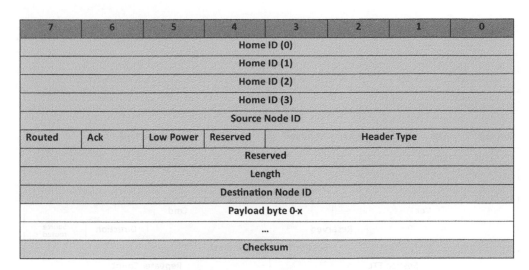

7	6	5	4	3	2	1	0
Home ID (0)							
Home ID (1)							
Home ID (2)							
Home ID (3)							
Source Node ID							
Routed	Ack	Low Power	Reserved	Header Type			
Reserved							
Length							
Destination Node ID							
Payload byte 0-x							
...							
Checksum							

Figure 8.1   Z-wave singlecast frame format.

Z-wave frames begin with a preamble, and encapsulate the transmitted data between a start of frame marker and an end of frame marker. The data itself is Manchester encoded at 9.6 kbps, and NRZ encoded at 40 kbps.

The collision-avoidance mechanism is based on delaying transmissions randomly after the last occurrence of RF activity on the channel.

8.2.4   Transfer Layer

The Z-wave transfer layer controls the transmitting and receiving of frames.

The general format of a Z-wave datagram transfer layer frame is illustrated in Figure 8.1.

Singlecast **frames** are sent to one specific node ID, and can optionally include a route (list of slave nodes to the destination). A flag indicates if an acknowledgment is desired or not (Acks are singlecast frames with no data, and generated hop by hop, with some optimization mechanism). Singlecasts get retransmitted if no Ack is received.

A *Multicast* **frame** can target a range of selected nodes (1 to 232). The format of this frame is identical to that of a unicast frame, except that the destination node ID is replaced by an address offset and one or more of mask bytes. The receiving nodes do not acknowledge a multicast frame. If reliable communication is required, a multicast frame must be followed by a singlecast frame.

A *Broadcast* **frame** targets all nodes in the Z-wave network with the specified Home ID and is not acknowledged. It is used for instance to transmit the *node information frame* when a node action button is pressed. The format of this frame is identical to that of a unicast frame, except that the destination node ID is set to FFh (255).

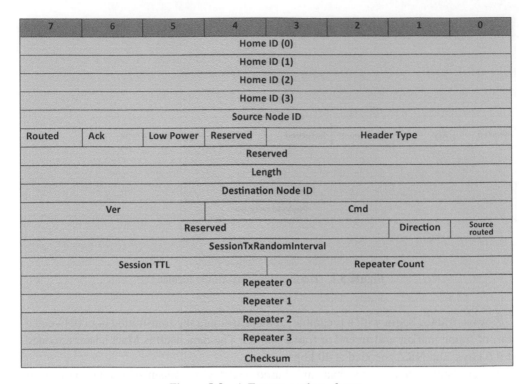

7	6	5	4	3	2	1	0
Home ID (0)							
Home ID (1)							
Home ID (2)							
Home ID (3)							
Source Node ID							
Routed	Ack	Low Power	Reserved	Header Type			
Reserved							
Length							
Destination Node ID							
Ver			Cmd				
Reserved					Direction		Source routed
SessionTxRandomInterval							
Session TTL			Repeater Count				
Repeater 0							
Repeater 1							
Repeater 2							
Repeater 3							
Checksum							

Figure 8.2 A Z-wave explorer frame.

An *Explorer* **frame** (Figure 8.2) is a special class of the broadcast frame. All nodes in the direct range of the originator receive an explorer frame. The handling of an explorer frame then depends on the destination address (a single node or all nodes), and a number of configuration flags, which determine:

– whether the frame should be forwarded by the receiving node specifically, or all nodes that receive it;
– whether frames with identical source node ID and sequence number should be discarded.

A *SearchRequest* addresses a particular node but must be forwarded by all nodes (network flooding). A *SearchResult* addresses a particular node and must be forwarded only by all nodes in a route. A *SearchStop* addresses a particular node and must be forwarded only by all nodes in a route but all nodes must discard frames resembling the frame (identified by srcNodeId + seqNo), in order to stop the network flooding initiated by a *SearchRequest*.

An explorer frame can carry a command to a destination NodeID through a list of repeaters, and is used for autoinclusion (network wide inclusion) and route resolution. The SessionTTL parameter, initially set to 4, is decremented by each forwarding node, ensuring that the flooding process terminates.

8.2.4.1   Battery-Powered Nodes

Battery-powered nodes pose a specific problem, because they are awake only periodically, and the device timers are not accurate enough to maintain any form of synchronization that would allow the sender of a frame to "guess" when to send it.

In a Z-wave network, sending configuration information to a routing slave that can enter sleep mode involves polling: the routing slave can use a "dial up" service to for example, signal to a static controller that they are ready to receive new configuration information. For this purpose the *wake-up command class* is used. Such nodes must know the route to the static controller (these routes are called *return routes*), and periodically issue a *wake up notification* command to any always listening device (it may be broadcast if the target node ID has not been configured by the *wake up interval set* command), requesting for pending commands. The target node will send any pending commands, then issue a *wake up no more information* command so that the battery-powered node can go back to sleep.

This communication method implies a relatively high latency, which is not acceptable to all applications. An example of such an application is a network of smoke detectors, which are all required to sound an alarm signal as soon as one detector has been triggered. For this type of application Zensys introduced the notion of ZensorNet™ routing slaves.

ZensorNet™ routing slaves use a specific a specific link layer mechanism. They use a wake-up beam, ZensorNet binding and ZensorNet flooding (no routing). When a ZensorNet™ device needs to send a frame to a network of ZensorNet™ routing slaves, it uses a wake-up beam longer than the configured wake up interval of devices in the network. This ensures that all other ZensorNet™ devices will receive the wake-up beam. After receiving such a beam, each ZensorNet™ devices starts beaming its own wake-up beam, flooding the network. Each device beams a given frame only once, preventing flooding loops.

Up to 16 ZensorNet™ routing slave nodes can exist in a ZensorNet™ network.

8.2.5   Routing Layer

The routing layer controls the routing of frames in the network. Z-wave networks use a source-routing mechanism: the initiator of a frame generates a complete route to the end destination through a number of repeaters.

The route consists in a sequence of node IDs that is placed in the frame, as illustrated in Figure 8.3. The frame becomes a *routed singlecast* (Figure 8.3). An always listening node receiving a *routed singlecast* with its node ID at the top of the repeater's list will repeat this frame, just removing its node ID from the list of repeaters. This ensures that routing loops cannot occur.

A routed acknowledge frame has the same structure, but the Ack flag is set and it carries no data in it.

If a device has tried all configured routes to a target without success, it issues a special "SearchRequest" explorer frame for the target node ID, embedding the command it was

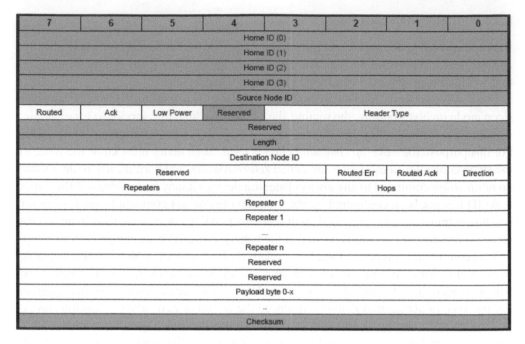

7	6	5	4	3	2	1	0
Home ID (0)							
Home ID (1)							
Home ID (2)							
Home ID (3)							
Source Node ID							
Routed	Ack	Low Power	Reserved	Header Type			
Reserved							
Length							
Destination Node ID							
Reserved					Routed Err	Routed Ack	Direction
Repeaters				Hops			
Repeater 0							
Repeater 1							
...							
Repeater n							
Reserved							
Reserved							
Payload byte 0-x							
..							
Checksum							

Figure 8.3 Z-wave routed singlecast frame.

trying to send. All slave nodes that receive this frame forward a copy, and hopefully the frame finally reaches the target device. The command is executed only once even if multiple copies reach the destination, because they all have the same sequence number. The target device then sends a "SearchReply" back along the route that was just discovered, and the initiating device learns this new route. A SearchStop explorer frame is sent immediately, ensuring that all pending search frames are killed, stopping network flooding as soon as possible.

8.2.5.1 Overview of Z-Wave Addressing: Home ID, Node ID, Primary and Secondary Controller, Inclusion

Z-wave uses a 32-bit *Home ID* identifier to separate different Z-wave networks (e.g., two adjacent houses). Each device is allocated an 8-bit *node ID*, unique for a given network identified by its *home ID*. A Z-wave network can have up to 232 nodes for any given *home ID*.

The *home ID* is configured[2] in each Z-wave *primary controller*. During the *inclusion* process the Z-wave *primary controller* allocates a unique *node ID* to each Z-wave device, and also configures the *home ID* allocated to each device.

---

[2] Generated randomly and renewed after each controller exclusion, for Zensys SDK 4.5 and above.

When a *controller* is included in an existing Z-wave network (controlled by a *primary controller*), it becomes a *secondary controller*, and is assigned a unique *Node ID* and the same *home ID* as the *primary controller*. In addition, for battery-powered nodes, the primary controller assigns a return route to the device (see Section 8.2.4.1)

8.2.5.2 Inclusion

With most current Z-wave networks, the inclusion process is manual. An action must be taken on the controller to enter a waiting mode, then a button must be pressed on the Z-ware peripheral to be included (one peripheral at a time). The peripheral sends a *node information frame* (NIF) to the controller, which must be reachable directly by the device during the association process. The NIF specifies:

- The *basic device class* that defines the protocol library used (portable controller, static controller, slave, and routing slave).
- The *device class*, for example, generic controller, static controller, binary switch, multilevel switch, binary sensor, multilevel sensor . . . . The device class describes the type and functionality of a device, and implies mandatory command classes that must be supported. The generic device class defines the broad functionality of a device (e.g., multilevel switch), and the specific device class defines more precisely the function of the device (e.g., multilevel power switch for a light dimmer, motor control class for automated window shades, both of generic device class multilevel power switch).
- All *command classes* supported by the device (except secure commands, see security).

The controller replies by assigning a Node ID to the device, and formally ending the transfer.

On recent Z-wave networks supporting the *"explorer frame"* (see Figure 8.2), the peripheral being associated does not need to be reachable directly by the controller during the association process: the association request and subsequent controller configuration commands can be routed through other nodes already part of the same network. Zensys calls this the *"autoinclusion"*, because devices supporting this feature enter "autoassociation mode" automatically on power-up. A periodic association request is sent (after increasing random intervals in order to minimize collisions) via explorer frames. The explorer frame floods the network, via adjacent nodes, with a TTL that prevents loops, trying to reach a static controller. When the static controller finally receives the request, it returns a routed response.

Immediately after inclusion, a node or secondary controller is requested by the controller to look for its neighbor nodes on the network (by sending NOP frames and listening for ACKs), and reports the node IDs of the devices within range back to the controller.

For any given node N (and secondary controllers), the primary controller knows which nodes are reachable by N (the node neighbor list). From this information it builds its own routing table, and sends all relevant information to the secondary controller to build its local routing table:

- The node table (node IDs as well as device/command classes);
- The node neighbor list;
- The SUC ID server (see controllers), if any.

Optionally, the controller can exchange information about groups (names and nodes) and scenes (names, nodes and levels in nodes), see application layer for details.

8.2.5.3   Network Management

When an "always listening" node is moved or becomes unreachable with the current routing configuration due a changing RF environment, Primary controllers or controllers with SUC/SIS functionality can initiate a "rediscovery" procedure to get new routes to the moved nodes, healing the network. A similar rediscovery procedure exists for "not listening" nodes.

As the node ID is allocated by the network controller, in case a node fails, it is possible to replace that node while preserving the node ID.

8.2.6   Application Layer

8.2.6.1   Commands

The Z-wave protocol supports the applications by defining a standard command structure, ensuring interoperability. Each command encapsulated in a Z-wave frame is composed of:

- An 8-bit command class identifier. Command classes are defined for the Z-wave protocol and for Z-wave applications, for example, COMMAND_CLASS_BATTERY.
- An 8-bit command identifier (an extension mechanism ensures that up to 4000 commands can be defined).
- A list of command parameters. For any command, the list of parameters can be extended as new versions are released. Devices supporting older version truncate the list of parameters to just those that they can understand. Devices can be queried for the command class version that they support.

Devices declare the supported command classes during the inclusion process, as part of the *node information frame* that is sent as a broadcast any time the node action button is pressed or as a response to a controller "*get node information*" command.

Most devices support the *basic command class*. Through primitives such as BASIC_SET or BASIC_SET, it is possible to set or read values for simple devices (like dimmers, power plugs or thermometers) which do not require a more complex command syntax.

Most devices will support additional command classes besides the basic command class, for example, COMMAND_CLASS_BATTERY for battery-powered devices, COMMAND_CLASS_SENSOR_MULTILEVEL for multilevel sensors (e.g., temperature, luminosity, hygrometry sensors)

For some uses, the list of commands already defined might be insufficient: an extension mechanism is defined for proprietary commands. Zensys assigns manufacturer IDs to ensure that proprietary commands from distinct manufacturers do not interfere.

8.2.6.2 Multi-Instance Devices

Some devices implement several functions, for example, multiple relays, multiple switches, or sensors for multiple physical parameters (e.g., temperature, humidity, pressure). As a consequence, it may not be enough to send a command to a node ID to identify the exact action desired. Z-wave solves that issue with the concept of multi-instance devices (MI). Multi-instance devices advertise support for the *Multi-Instance* command class in the *network information frame*.

The *multi-instance command encapsulation* command is used to encapsulate commands sent to a Z-Wave node, and simply adds an instance parameter to the encapsulated command.

8.2.6.3 Associations

Most home area networking and fieldbus protocols define a form of "publish/subscribe" model. This intent of this feature is to allow devices to interact directly without requiring services from a central node: this improves the reactivity of the system (e.g., the time it takes to switch on a lamp after operating the switch), and the resistance to failures of the central controller.

Z-wave implements the "publish/subscribe" model through the concept of *associations*. Z-wave calls an "association" the operation by which a Z-wave device is configured to control another device or set of devices. For instance, a switch or movement sensor can control a lamp or group of lamps.

For each type of event that a controlling device can generate (e.g., "switch on event"), it should allocate a locally unique 8-bit *grouping* ID. Up to 255 types of events can be managed per device.

Associations are handled at the application level of the device code, not by the Zensys libraries. Associations are supported by two command classes:

- COMMAND_CLASS_ASSOCIATION: The *association set* command is used to add nodes, identified by their node IDs, to a given grouping identifier. The *multi-instance association set* command can be used for multi-instance devices (both a node Id and an Instance Id are provided). Routing slaves must also be configured by a controller with return routes in order to be able to reach the target nodes. Additional commands allow the controller to retrieve the association status of the nodes, remove associations, request the number of groupings, and so on.

• COMMAND_CLASS_ASSOCIATION_COMMAND_CONFIGURATION defines the commands necessary to configure which commands should be sent to a node part of a grouping ID when the corresponding event occurs. A *command record* consists of the grouping identifier, the Node ID and the complete command. The *command configuration set* command is used to specify which command should be sent to a specific node ID within a given grouping identifier. Additional commands allow retrieval of the maximum number of command records, the remaining number of free command records, and so on.

Both primary and secondary controllers can configure groupings and the required return routes. Ordinary nodes can setup command records in other nodes if they support commands from the COMMAND_CLASS_ASSOCIATION_COMMAND_ CONFIGURATION class.

8.2.6.4   Scenes

Z-wave *scenes* provide the toolbox that facilitates the simultaneous configuration of multiple devices, with minimal network activity and avoiding the flicker effects that might appear if individual configuration commands were sent to each individual device. This can be used, for instance, to preset lamp levels for common situations, for example, "home theater", "dinner", "chat around the fireplace" configurations.

Scene settings can be configured in devices using commands of defined by the scene actuator configuration command class (management of settings associated to a given scene ID), and activated by means of the commands defined in the scene activation command class.

Several scenes can be configured in a controller. Each scene defines a list of node IDs part of this scene, and an 8-bit level parameter (a value that will have the same effect on devices as if it was passed through a basic set command).

Once a scene has been configured, it can be activated using the SCENE_ACTIVATION_SET command (which can be multicast), specifying a scene ID and a dimming duration.

8.2.6.5   Security

Z-wave network security is optional, and implemented by commands defined in the *security command class*. When security is provided, for example, for door locks, it is based on a network-wide secret key (created by the primary controller at startup) which is used to encrypt *security-encapsulated secure commands*.

A device willing to send a *security-encapsulated secure command* to another starts by requesting a nonce from the target Node ID (using a nonce get, nonce report message exchange). The value of this nonce, as well as a source generated initialization vector, will be included in the security-encapsulated secure command, ensuring replay prevention. A message authentication code is also added to the message to prevent tampering. The destination node ID is sent in clear as part of a standard singlecast, therefore secure messages can be routed through nonsecure nodes.

AES 128 is used for authentication (message authentication code based on a Davies–Meyer hash) and encryption. The current SoC generation requires a software implementation. Hardware support for security has been announced for the fourth generation of the chip, ZW0401, which will implement AES-128 in hardware.

The network secret key is distributed during initial installation: right after the inclusion of a secure node, the primary controller sends *key set* security encapsulated commands to the secure node, using a temporary key (all zeroes).[3]

Developers decide, at the application level, which commands should be supported securely, and declare these commands to the controller separately (security commands supported get/report). All commands declared through the standard *node information frame* (NIF) are nonsecure.

---

[3] Future implementations may use a PIN-code-based encryption during the initial network key exchange.

Part Three

Legacy M2M Protocols for Utility Metering

Part Three

Legacy M2M Protocols for Utility Metering

9

M-Bus and Wireless M-Bus

9.1 Development of the Standard

The M-Bus standard was the result of collaboration between Dr. Horst Ziegler of the University of Paderborn, a chip maker (Texas Instruments) and a company focused on metering data management (Techem). In 1990 there was no established communication standard for the reading of utility metering devices: such a standard had to be low cost and adapted to battery-powered meters.

The original design uses a simple two-wire serial communication bus, and is documented at http://www.m-bus.com/. One major advantage of the new bus was that all meters and the reading device could be connected to the same wire (which is why it is called a bus).

The link layer used by M-Bus was initially standardized in 1990 as IEC 870-5-1 (Telecontrol Equipment and Systems/Transmission Protocols/Transmission Frame Formats) and IEC 870-5-2 (Link Transmission Procedures, 1992). The first standard to be published related to the M-Bus application layer was EN 1434 in 1997, where parts -2 and -3 define an application layer for a wire communication protocol dedicated to heat meters.

The standardization work is now managed by Cenelec Technical committee TC 294, which generalized the use of M-Bus for any type of meter readout in the EN 13 757 series:

- EN13757-1:2002 Data exchange (DLMS);
- EN13757-2:2004 Physical and link layer (M-Bus);
- EN13757-3:2004 Dedicated application layer (M-Bus);
- EN13757-4:2005 Wireless meter readout (wM-Bus);
- EN13757-5:2007 Relaying (network aspects);
- EN13757-6:2007 Data exchange (local bus).

The Internet of Things: Key Applications and Protocols, First Edition.
Olivier Hersent, David Boswarthick and Omar Elloumi.
© 2012 John Wiley & Sons, Ltd. Published 2012 by John Wiley & Sons, Ltd.

Table 9.1   M-Bus wired data transmission, physical layer bit encoding

	Master transmission	Slave transmission
Bit 0	+36 V	<1.5 mA
Bit 1	+24 V	11–20 mA

EN 13 757-2 defines the physical and link layer for the wired bus, while EN 13 757-3 defines the application layer (actual messages). The wireless version of M-Bus is defined in EN 13 757-4.

9.2   M-Bus Architecture

M-Bus was not originally designed for complex multihop networking, and it uses a simplified 4-layer model.

9.2.1   Physical Layer

The two-wire bus interconnects one master and several slaves, using asynchronous half-duplex serial data transmission. Slaves can be powered from the bus.

The nominal voltage of the bus is 36 V. Serial data transmission from the master to slaves uses bus voltage level shifts, while data transmission from the slaves to the master uses a modulation of the slave current consumption (Table 9.1).

The quiescent state is always a 1. The start bit of the serial transmission is therefore a "0", followed by 8 data bits, followed by a parity bit and a stop bit (always 1).

The communication speed can vary between 300 and 9600 bits/s, and up to 250 slaves can be connected over a single twisted pair, which can span up to 1000 m, depending on the number of slaves.

9.2.2   Link Layer

The data link layer is based on IEC 870-5, which defines several frame formats depending on the level of integrity protection required. M-Bus uses frame format class FT 1.2, and defines 4 types of frames (Figure 9.1).

The **single-character frame** is used to acknowledge transmissions.

The **short frame** has a fixed size.

The **long frame** has a length field (user data bytes + 3) that is transmitted twice, followed again by the start character.

The **control frame** is a long frame without user data.

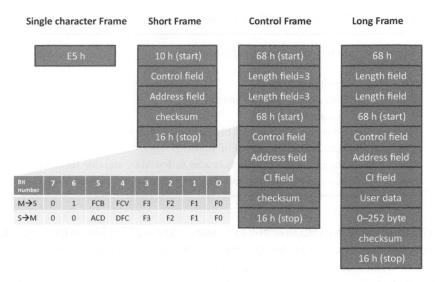

Figure 9.1 M-Bus frames.

Primary addresses 1 to 250 can be allocated to slaves. Address 0 is reserved for unconfigured slaves. Address 253 is reserved to indicate that the destination address has been set at the network layer. Addresses 254/255 are reserved for managing the local physical layer.

The bits of the control field (C) have a different use depending on the direction of the communication. The FCB bit is used as a one-bit frame counter (used if FCV bit is set). The DFC bit is used by slaves to signal that it can accept no further data. The ACD bit is set when a slave wants to transmit higher-priority data.

The CI field (control information) is part of the application layer.

The link layer defines two transmission services:

- Send/Confirm (SND/CNF): the response is a single-character frame. SND_NKE is used to start or restart a communication (the next master frame will have NCB = 1). SND_UD (optional) is used by the master to send user data to the slave.
- Request/Response (REQ/RSP): REQ_UD2 is used by the master to request data from the slave, and the slave can transfer its data to the slave using RSP_UD.

9.2.3 Network Layer

The M-Bus network is structured into zones, which consists of one or more segments interconnected by repeaters.

The standard provides a way to switch a given slave in a "selected" state, without using or requiring a primary address. In order to achieve this, the master sends a special

Table 9.2 M-Bus, common CI field values

00h-4Fh 54h-58h	Reserved for DLMS applications
50h	Application reset
51h	Data send (master to slave)
5Ch	Synchronize action
70h	Slave to master: report of application errors
71h	Slave to master: report of alarms
72h	Slave to master: 12-byte header followed by variable format data
78h	Slave to master: variable data format response without header

SND_UD message with CI=52h or 56h to primary address 253 (FDh), which specifies the secondary address of the slave (identification number, manufacturer, version and medium, specified or wildcarded). All slaves receiving this message compare these information elements with their own, the slave that matches, if any, enters "selected" state and sends a single-character response.

Subsequent requests to the selected slave(s) may be sent by SND_UD messages addressed to primary address FDh. Slaves remain in selected state until they receive a new selection command with a nonmatching secondary address.

The selection procedure, because it allows wildcarded parameters for individual address digits composing the secondary address, can also be used by masters for network discovery.

9.2.4 Application Layer

Control information (CI) field values indicate the formatting of the rest of the application data payload. Table 9.2 lists common values of the CI field.

Example user data formatting for CI=72h:

- A 12-byte header shown in Table 9.3.
- Variable data blocks: composed of one or more data records. Each data record is composed of a data record header (DRH), followed by the data itself.

   The DRH describes the data:
   - The data information block (DIB) of the DRH describes the length, type and coding of the data (e.g., 6-digit BCD, variable length 8-bit text string) as well as the storage number of the data concerned (register identifier of the value being read), and whether the value being read is an instant value, minimum value or invalid.
   - The value information block (VIB) contains the value of the unit and the multiplier. For instance a value information field of "E000 0nnn" corresponds to an energy measurement in Wh from 0,001 Wh to 10 000 Wh with a multiplier of $10^{(nnn-3)}$. See Table 9.4 for common unit codes.

Table 9.3 Application data header example

Ident. Nr	Manufacturer	Version	Device type	Access No	Status	Signature
Serial number allocated during manufacture. 8 BCD digits	Derived from 3-letter EN 61107 ASCII code of manufacturer ID	Determined by manufacturer		Response counter	Error flags, e.g., Power Low	Used for encryption
4 byte	2 byte	1 byte	1 byte	1 byte	1 byte	2 byte

Table 9.4 M-Bus value information field unit codes

Coding	Description	Range Coding	Range
E000 0nnn	Energy	$10(nnn-3)$ Wh	0.001 Wh to 10 000 Wh
E000 1nnn	Energy	$10(nnn)$ J	0.001 kJ to 10 000 kJ
E001 0nnn	Volume	$10(nnn-6)$ m^3	0.001 l to 10 000 l
E001 1nnn	Mass	$10(nnn-3)$ kg	0.001 kg to 10 000 kg
E010 00nn	On Time	nn = 00 s nn = 01 min nn = 10 h nn = 11 day	
E010 01nn	Operating Time	*as above*	
E010 1nnn	Power	$10(nnn-3)$ W	0.001 W to 10 000 W
E011 0nnn	Power	$10(nnn)$ J/h	0.001 kJ/h to 10 000 kJ/h
E011 1nnn	Volume Flow	$10(nnn-6)$ m^3/h	0.001 l/h to 10 000 l/h
E100 0nnn	Volume Flow ext.	$10(nnn-7)$ m^3/min	0.0001 l/min to 1000 l/min
E100 1nnn	Volume Flow ext.	$10(nnn-9)$ m^3/s	0.001 ml/s to 10 000 ml/s
E101 0nnn	Mass flow	$10(nnn-3)$ kg/h	0.001 kg/h to 10 000 kg/h
E101 10nn	Flow Temperature	$10(nn-3)$ °C	0.001 °C to 1 °C
E101 11nn	Return Temperature	$10(nn-3)$ °C	0.001 °C to 1 °C
E110 00nn	Temperature Difference	$10(nn-3)$ K	1 mK to 1000 mK
E110 01nn	External Temperature	$10(nn-3)$ °C	0.001 °C to 1 °C
E110 10nn	Pressure	$10(nn-3)$ bar	1 mbar to 1000 mbar
E110 110n	Time Point	$n = 0$ date $n = 1$ time and date	IEC 870-5-4 CP16 8-bit encoded date IEC 870-5-4 CP32 32-bit encoded date
E110 1110			dimensionless

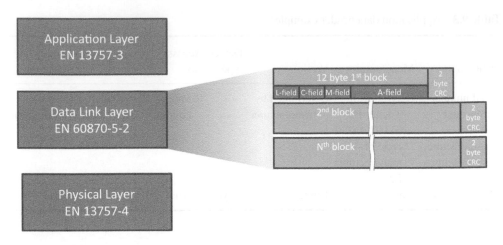

Figure 9.2   The wireless M-Bus protocol layers, details of the link layer.

9.3   Wireless M-Bus

Wireless M-Bus is specified EN13757-4:2005 (wireless meter readout) and uses the 868 MHz frequency, a lower frequency (169 MHz) enabling better penetration within buildings is being standardized as well (prEN13757-4:2011 Mode N). The application messages are specified in EN 13 757-3 as for the wired M-Bus.

The wireless M-Bus specification defines communications between a "meter" device and a "other" device, which usually corresponds to a data concentrator.

Wireless M-Bus uses the 3-layer IEC model, outlined in Figure 9.2.

9.3.1   Physical Layer

The physical layer is described by EN 13 757-4 "Communication system for meters and remote reading of meters, Part 4: Wireless meter readout, Radio meter reading for operation in the 868 to 870 MHz SRD band".

Several data transmission modes have been defined on the 868 MHz license free ISM band defined in Europe (Table 9.5): the stationary mode (S-mode), the frequent-transmit mode (T-mode) and the frequent-receive mode (R-mode). The physical and data link layers of the S-mode have been designed jointly with the KNX association, and are identical to KNX-RF (see Chapter 6). More recently, the N-mode, at 169MHZ, has been defined.

All modes have been designed to optimize the battery lifetime, the meter remains in sleep mode except during short transmission periods. Some modes enable bidirectional communication: the meter switches to receive mode for a short period after transmission. Even in bidirectional mode, the "other" side can never initiate a transmission, but needs to wait until the meter wakes up, transmits at least a message, and then switches to receive mode.

Table 9.5 wM-Bus transmission modes

Mode	Data rate from meter (chip rate)	Direction	
R2	4.8 kbps	Meter → Other	
S1/S1m	32 768 kbps	Meter → Other	868.3 MHz (600 kHz channel), 2FSK modulation.
S2		Meter ↔ Other	
T1	100 kbps	Meter → Other	
T2		Meter ↔ Other	868.95 MHz for other to meter, 500 kHz channel.

Mode T provides the shortest transmission time and the longest battery life for a wireless meter. The battery-life requirements are typically 10 to 20 years.

Wireless M-Bus bit encoding use a higher frequency for encoding chip "1" and a lower frequency for encoding chip "0". Transmissions begin by a preamble composed of an integer number of "01", followed by a synchronization word.

For data, the wireless M-Bus specification uses two channel-encoding methods:

- Manchester encoding: data bit "1" is encoded by chip pattern "01", while bit "0" is encoded by chip pattern "10". This is a classic encoding that was used to facilitate clock recovery, although with most modern radio chips, it would no longer be required. The resulting data rate is half of the chip rate.
- Three out of six encoding: each byte of data is split in nibbles of 4 bits, each 4-bit nibble is then encoded as 6 chips. As a result, each data byte is encoded in one and a half chip bytes.

The encoded data is then followed by a trailer of two to eight chips.

The various physical layer parameters used for each transmission mode are summarized in Table 9.6.

Table 9.6 Wireless M-Bus physical layer parameters

Mode	Preamble (chips)	Sync word (chips)	Encoding
R2	78	18	Manchester
S1	558 (about 17 ms !)	18	Manchester
S2 (two-way)	30	18	Manchester
T 1(one-way)	38	10	3 out of 6 encoding
T2			
other to meter			Manchester

9.3.2 Data-Link Layer

The data-link layer is defined in IEC 60 870-2-1:1992, Telecontrol equipment and systems, Part 5: Transmission protocols, Section 1: Link transmission procedure, and IEC 60 870-1-1:1990, Telecontrol equipment and systems, Part 5: Transmission protocols, Section 1: Transmission frame formats.

The data-link layer manages framing and CRC generation and detection, provides physical addressing and manages acknowledges (bidirectional modes only).

The wireless M-Bus frame format used in EN 13 757-4:2005 is derived from the frame type 3 (FT3) defined in EC60870-5-2. The frame consists of one or more blocks of data, as illustrated in Figure 9.2. Each block includes a 16-bit CRC field. The first block is a fixed length block of 12 bytes that includes the L-field, C-field, M-field, and A-Field.

- L-Field: in variable packet size mode, the transmission frame format begins by a length indication (L-field), which indicates the number of link-layer payload bytes not including the L, M and A fields, CRC bytes or encoding, from 0 to 255. The L-field itself is encoded using Manchester or 3 out of 6 encoding. Due to the various overheads, the actual maximum value of the L field may be limited by the underlying radio chip packet handler.
- C-Field: the C-field identifies the frame type: SEND, CONFIRM, REQUEST, or RESPOND. For SEND and REQUEST frames, the C-field also indicates whether a CONFIRM or RESPOND is expected.
- M-Field: the manufacturer's 3-letter code (encoded as 3×5 bits by taking the ASCII code and substracting 0x40). Assigned codes are listed at http://www.dlms.com/flag/INDEX.HTM
- A-Field: each device is assigned a unique 6-byte address, assigned by the manufacturer. As in the case of the wired M-Bus, only meters are assigned addresses: the address included in the frame is always the originator address for send and request frames, and the target address for confirm and response data frames. The "other" side remains anonymous.
- CI-Field: the application header specified the application payload type. The possible values are specified in EN13757-4:2005.
- CRC: a 2-byte CRC is added for each 16-byte block, using polynomial $x16 + x13 + x12 + x11 + x10 + x8 + x6 + x5 + x2 + 1$.

9.3.3 Application Layer

The application layer is defined in EN 13 757-3, Communication system for meters and remote reading of meters, Part 3: Dedicated application layer. It is identical to the application layer of the wired M-Bus, which facilitates bridged M-Bus applications connecting a wired side with a wireless side.

9.3.4   Security

M-Bus data can optionally be encrypted. The original specification used the DES algorithm, while more recent specifications now use the AES 128 algorithm.

When using encryption, the meter is configured with a key, which is shared by the master. Some headers are always sent unencrypted (e.g., the header of Table 9.3), and a portion of the header indicates the encryption method (high byte of the signature field in the example of Table 9.3, for instance DES cipher block chaining). The encrypted data follows.

10

The ANSI C12 Suite

Jean-Marc Ballot
Alcatel-Lucent

10.1 Introduction

Before the publication of C12, meters from different vendors in the USA did not use the same data formats or communication protocols. C12 is a standard suite specified by the American National Standards Institute (ANSI), which provides an interoperable solution for data formats, data structures, and communication protocols used in automatic metering infrastructure (AMI) projects. Although this standard is focused on the American national market, C.12 is used in smart metering systems of many countries (in particular its variant OSGP, deployed by Echelon, Inc).

The standard data structure is specified in the ANSI C12.19 document. It is defined as a set of tables. When these tables share a common purpose or they are relative to a common feature of the meter, then the tables are included in a specific chapter called a "decade".

The first communication protocol that made use of the C12.19 tables was specified in the ANSI C12.18 document. The first release was published in 1996 and describes the communications between a C12.18 meter and a C12.18 client by means of an optical port. A C12.18 client could be a hand-held reader, a laptop, or any device with an optical port.

In 1999, ANSI C12.21 was specified for communications between a C12 device and C12 client via a modem. This offered a first solution for AMI projects.

In 2007, a new specification took into account the development of data networks not using modems. ANSI C12.22 is a "Protocol Specification for Interfacing to Data Communication Networks". The goal of this new item of the C12 suite is to allow interactions with C12.19 table data over any networking communications system.

The Internet of Things: Key Applications and Protocols, First Edition.
Olivier Hersent, David Boswarthick and Omar Elloumi.
© 2012 John Wiley & Sons, Ltd. Published 2012 by John Wiley & Sons, Ltd.

Beyond the ANSI specifications, RFC 6142 "ANSI C12.22, IEEE 1703 and MC12.22 Transport Over IP", published in March 2011, proposes a framework for transporting ANSI C12.22 Application Layer messages on an IP network.

10.2   C12.19: The C12 Data Model

ANSI C12.19 is the "American National Standard for Utility Industry End Device Data Tables".

C12.19 defines a data structure used for representing metering data and metering functions exposed by a metering equipment to a client machine. C12.19 does not contain any protocol for the transport of the data, only the data structure is specified. As briefly mentioned in the introduction, the data structure is defined as a set of standard tables. When these standard tables share a common purpose or they are relative to a common feature of the meter, then the tables are included in a specific chapter called a "decade". Each decade covers a specific area of functionality. The version of ANSI C12.19 published in 2007 contains 17 decades (the original version published in the 1990s contained fewer decades than the current version). Table 10.1 provides the list of C12.19 decades:

Beyond these standard tables, ANSI C12.19 also provides a standard way to add proprietary tables. These tables are called manufacturer tables. If they follow the general

Table 10.1   C12.19 table decades

Decade number	Name of the Decade	Number of Tables in the Decade
0	Configuration Tables	9
1	Data Source Tables	9
2	Register Tables	9
3	Local Display Tables	5
4	Security Tables	7
5	Time-of-Use Tables	7
6	Load Profile Tables	8
7	History & Events Logs	10
8	User-Defined Tables	10
9	Telephone Control Tables	9
10	Extended Source Tables	4
11	Load Control & Pricing Tables	9
12	Network Control Tables	Temporarily defined in ANSI C12.22
13	Relay Control Tables	Temporarily defined in ANSI C12.22
14	Extended User Defined Tables	4
15	Quality of Service Tables	9
16	One-Way Tables	5

rules for the table format, it is possible for a manufacturer to introduce some value-added functions in its products.

ANSI C12.19 carefully defines all the data types that are used in the definition of the data structure. The communication protocol is not described, only the data format is specified.

The transport of table structures is not specified by C12.19 but it is mentioned that this transport *"is dependent only on the presence of basic Read and Write services (e.g., those as defined in ANSI C12.18, ANSI C12.21 and ANSI C12.22)"* (extract from C12.19 section 8). The design of C12.19 is natively "RESTful" (see Chapter 13)! The structure of the tables ensures that any operation required to manipulate the C12.19 tables can be performed only with the basic read and write services. ANSI C12.19 provides some basic requirements for the read and write services, which must be implemented by all C12 devices, but manufacturers may additionally implement more primitives to interact with C12.19 tables.

10.2.1   The Read and Write Minimum Services

- The read service request allows the transfer of table data from a sending party to a receiving party. The read service can be used for full table read or a partial table read. In the case of full transfer, only the Table_Identifier is provided in the read request. In the case of partial table read, some additional optional parameters have to be provided in order to choose the records and record fields that are requested. Two addressing methods are specified for a partial read:
  - a first method by providing up to 5 indexes relative to the table and optionally an element count starting at the indexed position;
  - a second one by providing an offset (in octets) relative to the beginning of the table and optionally an octet count starting at the indicated position.

If the end device that receives the read request does not support the method, the entire table is retrieved.

- The write service allows nonsolicited data to be sent to a receiving party. As for the read service, the write service request allows a complete or a partial table write to be performed, and the partial table write service can use indices or an offset.

Besides the read and write service, C12.19 supports 27 standard commands, and even manufacturer-defined commands, but the implementation of commands only uses the basic read and write services, using special-purpose tables (Tables 07 and 08). C12.19 does not compromise with the REST design!

10.2.2   Some Remarkable C12.19 Tables

The first table of the first decade, Table 00, plays a special role. It is called GEN_CONFIG_TBL and contains all the information relative to the configuration of the end device. For example, it contains the full list of supported tables and procedures.

Tables 07 and 08 in Decade 0 also play a specific role. They are called "Table 07: Procedure Initiate Table" and "Table 08: Procedure Response Table" and are used for enabling the execution of commands. When an initiator wants to request the execution of a command in a meter, it has to write in Table 07 some parameters that explicitly provide information about the procedure to be executed. The command response, that is, the result of the procedure execution, is placed in Table 08 in order to be read by the initiator of the command.

It has to be noted that these two special-purpose Tables 07 and 08 are not able to buffer commands. They enable execution of only one command at a time: the specification explicitly mentions that *"If a procedure initiate request is followed by another procedure initiate request, the procedure response for the first procedure initiate request may be lost"*.

The list of procedures that may be executed by using Tables 07 and 08 contains 28 standard procedures. Among the 28 procedures we can mention:

- cold start;
- warm start;
- save configuration;
- remote reset;
- set date and/or time;
- execute diagnostics.

10.3   C12.18: Basic Point-to-Point Communication Over an Optical Port

ANSI C12.18 was the first standardized protocol that was specified to interact with ANSI C12.19 Data Tables. The first release was published in 1996 then revised in 2006. It describes the communications between an electric metering equipment and another device used as a client via an optical port. The client device is typically a hand-held reader or a laptop used for reading or writing the meter internal data.

ANSI C12.18 focuses on the physical, data link and application layers. Layers 3 to 6 of the OSI model are out of scope. The three main functional areas covered by C12.18 are the following:

- modification of the communication channel;
- transport of information to and from the metering device;
- closure of the communication channel when communications are complete.

The application layer defines the PSEM (protocol specifications for electric metering) language that provides basic services that are used for channel configuration and information retrieval. Each service uses a request–response scheme. Nine services are defined:

- Identification service: this is the first service that shall be invoked after the establishment of the physical connection. The version and revision numbers of the protocol are returned by the service.
- Read service: this is used for triggering the transfer of table data from the requested device to the requesting one. As mandated by ANSI C12.19, both complete and partial transfer are possible. The complete transfer is mandatory. The partial transfer may use one of the two possible options: index based or offset based.
- Write service: this is used to transfer a table data to a target device. As mandated by ANSI C12.19, both complete and partial transfer are possible. The complete transfer is mandatory. The partial transfer may use one of the two possible options: index based or offset based.
- Logon service: this is used to setup a session without establishing the access permissions yet. These permissions will be established later through the security service.
- Security service: this is used for establishing access permissions. It is based on the use of a password as a mean for selecting access permissions. This service cannot be invoked before the logon service because a session has to be established as a prerequisite. The received password is compared with the one stored in the password table of the security decade.
- Logoff service: this is used for terminating the session previously established via the logon service.
- Negotiate service: this is an optional service used to reconfigure the communication channel in the case of the desired communication parameters differ from the default values. Baud rate and packet size are among the negotiable parameters.
- Wait service: this is used for maintaining an already established communication channel beyond the time-out value that ensures automatic termination. The value of the time-out will be reset to the previous value once a valid packet is received.
- Terminate service: this is used for immediately interrupting the communication channel. Generally, this service is used in the case of excessive errors or security issues.

Besides these high-level application layer services, ANSI C12.18 also provides settings for Layer-2 and Layer-1 establishment. Baud rate, number of packets, packet size channel traffic time-out, data type, data format and data polarity are among the handled parameters.

10.4   C12.21: An Extension of C12.18 for Modem Communication

ANSI C12.21 "Protocol Specification for Telephone Modem Communication" is an extension to the C12.18 standard. C12.18 was the first standardized protocol allowing interaction with ANSI C12.19 tables, but it was still necessary to be in the immediate vicinity of the C12 Device when handling the tables. C12.21 allows remote interactions over a telephone network.

The three main area of functionalities already provided by C12.18 are not modified (i.e. modification of the communication channel, transport of information, and closure of the communication channel).

The PSEM (protocol specifications for electric metering) now contains 12 services instead of 9 in the C12.18 specification.

- 7 services are identical to those in C12.18: read, write, logon, security. logoff, nego-tiate, wait. Actually, logoff service is very slightly modified in the terminology of its description but not in its functionality.
- 2 services (identification and terminate service) are modified compared with their C12.18 versions.
  - Identification service: the modification implements basic negotiation of the authenti-cation algorithm. If authentication is supported, then the authentication itself will be performed by calling a new service of C12.21: the authenticate service (see below).
  - Terminate service: it is used for immediately returning the communication channel to its "base state" that is, the state in which the channel is still established but with the default parameters. In this state there is no established session.
- 3 new services are provided: timing setup, disconnect, and authenticate.
  - Timing_Setup service: this is an optional service that allows configuring some timers or number of retry attempts used in the communication channel establishment, when these values differ from the default values.
  - Disconnect service: this is used for immediately interrupting the communication channel. The disconnect service in the C12.21 is a redesign of the terminate service in the C12.18, main use case for this service is to interrupt the communication when too many errors or security issues are observed.
  - Authenticate service: this new service was required as a result of the modification of the identification service. The C12.21 identification service negotiates the authentication algorithm supported by the end device. After establishing a session with the logon service, the authenticate service will be used in order to perform mutual authentication at session level.

10.4.1   Interactions with the Data-Link Layer

They are quite limited. The communication channel of the modem is established with a set of default parameters. The service layer only has the possibility, after calling the identification service and before calling the logon service, to call either the negotiate service or the Timing_Setup service (or both) in order to modify packet size, packet number for reassembly, timers, or retry attempts number.

10.4.2   Modifications and Additions to C12.19 Tables

Beyond the modification of existing services or addition of new services, C12.21 also specifies some changes in the C12.19 tables. Some existing tables were modified and some new tables were added.

  Some of the most significant changes are listed below:

- C12.19 Table 07 (procedure initiate table) was modified in order to add a new standard procedure that did not existing in the original version of the C12.19. This new procedure triggers an immediate call establishment with a phone number specified as a procedure parameter.
- A new decade (no. 9) was added to the original C12.19. This decade contains 7 new tables associated with the use of a telephone modem.

10.5   C12.22: C12.19 Tables Transport Over Any Networking Communication System

ANSI C12.22 "Protocol Specification for Interfacing to Data Communication Networks" was made necessary by the development of new networking technologies. The approach of C12.18 that defines a communication protocol for a given network was no longer practical. C12.19 introduces new concepts enabling transport C12.19 data from meters to a back-end central system over any kind of communication network.

  ANSI C12.22 defines several types of network elements that are used in a reference topology. Interfaces between different types of network element are described in the standard. Some new data tables are also added to the ANSI C12.19-1997 standard as required by the new C12.22 interfaces. Some existing tables are also modified in order to ensure compatibility with the new C12.22 standard.

10.5.1   Reference Topology and Network Elements

The reference topology defined by C12.22 is outlined in Figure 10.1.

  This reference topology makes use of different types of network elements:

- C12.22 Host: this is a termination point in a C12.22 network. It may be an authentication host or/and notification host. An authentication host performs the authentication tasks for a registering node. A Notification Host is the applicative part that is able to interpret the C12.19 data structures and that needs to be notified when new nodes are registered.
- C12.22 Device: this is a network element that contains a C12.22 application (the C12.19 data structures, the associated protocol, and the control of associations). In order to enable communications between a C12.22 device and the C12.22 network, a

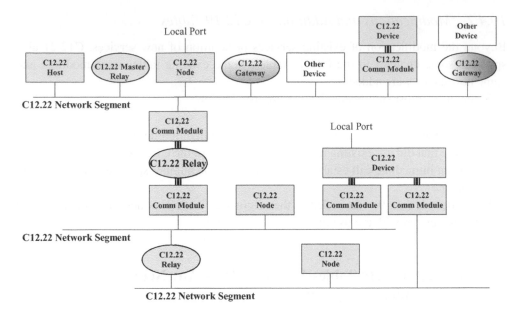

Figure 10.1   C12.22 reference topology (ANSI C12.22, Chapter 5, Figure 5.1).

C12.22 device has to implement one (at least) interface to a C12.22 communication
module.

- C12.22 Communication Module: this is a hardware device that implements the Layers 6
  to 1 of the OSI model allowing communications between a C12.22 Device and a C12.22
  network through an interface fully defined in the C12.22 specification.
- C12.22 Node: it is a combined C12.22 communication-module/device network element.
- C12.22 Master Relay: this is a network element in charge of receiving the registration
  requests coming from the C12.22 network elements that belong to its domain. It is also in
  charge of propagating the registration request to the appropriate C12.22 authentication
  host. A C12.22 master relay is on top of a hierarchical topology of C12.22 relays. A
  C12.22 master relay contains all the functionalities of a C12.22 Relay.
- C12.22 Relay: this is a network element that ensures the address translation between
  the Layer 7 address of a C12.22 node and its network address. This layer 7 address is
  called ApTitle (application process title). A C12.22 relay also uses the ApTitle in order
  to provide a C12.22 message-forwarding service when the lower layers do not provide
  such forwarding capability. In this case, when a message is sent to a C12.22 Relay
  with a called ApTitle different from the relay's ApTitle, then the C12.22 relay is in
  charge to forward the message to the final destination or to the first C12.22 relay in the
  path. A C12.22 relay maintains internal routing tables in order to provide this routing
  service.

- C12.22 Gateway: this is a protocol converter from the C12.22 protocol to any other protocol. It is used for enabling communications between a C12.22 node and a non-C12.22 node.

10.5.2   C12.22 Node to C12.22 Network Communications

The protocol stack used for communicating between a C12.22 node and a C12.22 network is only defined at layer 7 that is, at the application layer. The other layers (6 to 1) are "open to any network protocol". This application layer obviously contains the C12.19 data tables and also provides an evolution of the C12.21 PSEM (protocol specification for electric metering). This new version of the PSEM protocol contains 13 services:

- Three services are unchanged: the read, write and security services.
- Six services are modified (compared to C12.21): identification, logon, logoff, terminate, disconnect, and wait services.
- Four new services are provided: registration, deregistration, resolve, and trace services.
  - Registration service: this is used by a C12.22 network element in order to declare itself to the hierarchical structure of C12.22 relays. A C12.22 element has to send a registration request to a C12.22 master relay. During the initial registration a new routing table entry is added to all the C12.22 relays on the path to the master relay. Routing table entries are a soft state, and subsequent periodic registration requests and necessary to keep the routing table entry valid. After receiving a registration request, a C12.22 master relay has to forward it to the C12.22 authentication host.
  - Deregistration service: the effect of this service when it is called by a C12.22 network element is a deletion of the corresponding routing table entry from all the C12.22 master relay and relays. The removal of this network element is taken into account by all other C12.22 elements including authentication and notification hosts.
  - Resolve service: this enables communication between two C12.22 nodes that belong to the same local area network. When a requesting C12.22 node (X) needs to directly communicate with another local C12.22 node Y, it sends a resolve request with the ApTitle of node Y to its C12.22 relay in order to retrieve the native network address of node Y.
  - Trace service: when invoked by a C12.22 node, this service returns the path of C12.22 relays between the requesting node and a target C12.22 Node. The target node is not involved in the processing of the trace service that is only performed at the C12.22 relay level.

As in C12.18 and C12.21 standards, partial table access is possible in C12.22 specification by using one of the two defined methods: index or offset based.

An extended mode of the PSEM, called EPSEM (extended PSEM) is specified. EPSEM allows sending multiple requests and receiving multiple responses simultaneously.

In order to convey the APDUs (application protocol data units) that contain EPSEM services and their associated payload, the C12.22 standard uses the ACSE (association control service element) encoding method specified in ISO 8650-1. ACSE is an envelope for EPSEM primitives that also allows to transport association parameters and some security parameters when a secure transaction is required (C12.22 security mechanism supports both authentication and encryption).

10.5.3   C12.22 Device to C12.22 Communication Module Interface

In order to model the communication ports of C12.22 meters, C12.22 introduces the concept of C12.22 communication modules. A given meter may support multiple types of "plug-in" communication modules, using a standard connector and serial protocol.

As represented in Figure 10.2, a communication module is connected to the C12.22 Device through an interface that is fully defined in the C12.22 standard. It is also connected to any LAN (e.g., ZigBee, . . .), WAN (DSL, GPRS, . . .), or MAN (Ethernet, . . .). When a short-range connection is used, the communication module may communicate with a C12.22 relay that implements the same network technology.

The C12.22 device/C12.22 communication module protocol stack is fully defined in C12.22, and is outlined in Figure 10.3.

The application layer in a C12.22 device (and optionally in a C12.22 communication module) is identical to the one implemented in a C12.22 node and described in the previous section.

As mentioned in Figure 10.3, the transport layer specifies a set of 6 services that have the following functionalities:

* Negotiate service: this is used when a C12.22 communication module detects an attached C12.22 device. The service negotiates communication parameters settings with the C12.22 device. Typical examples of communication parameters that may be negotiated

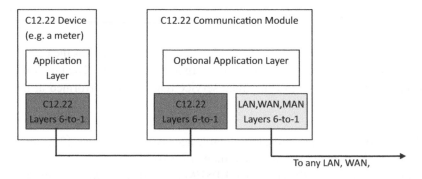

Figure 10.2   C12.22 Communication module.

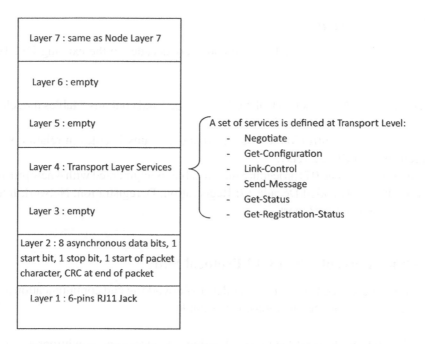

Figure 10.3   C12.22 Communication module to node protocol stack.

by invoking this service are baud-rate, maximum-packet-size, maximum-number-of-packet for reassembly function, . . .

- Get-Configuration service: this service is invoked by the communication module in order to request its configuration from the C12.22 Device. ApTitles, native address, or device class are among the exchanged parameters.
- Link-Control service: by using this service the C12.22 device can control communications from the communication module to the C12.22 network. It can enable or disable the communication interface, the registration process, direct communication with target nodes on the same network segment, and other functions.
- Send-Message service: this is the service enabling exchange of ACSE-PDUs. It is used by the C12.22 device when sending a message to a C12.22 communication module or by a communication module when transmitting a message received from the C12.22 network.
- Get-Status service: the C12.22 device may use this service in order to get information about the communication interface of the C12.22 communication module. The device may also request some network statistics about the communication status.
- Get-Registration-Status service: after a registration is performed, by invoking this service the C12.22 device can retrieve registration information from the communication module. The returned information includes relay address and ApTitle, registration period, and the amount of time left before the end of the current registration.

10.5.4   C12.19 Updates

C12.22 updates some existing tables and adds new decades to the existing C12.19 data tables:

- Decade 12 "Node Network Control Tables" is added and contains 8 tables modeling the C12.22 node access to a C12.22 network
- Decade 13 "Relay Control Tables" is added and contains 7 tables in relation with the management of a C12.22 relay.
- The content of the Table 07 (procedure initiate table) is augmented with 4 new procedures related to the newly added Decade 12: Registration, Deregistration, Network Interface Control, and Exception Report.

10.6   Other Parts of ANSI C12 Protocol Suite

The previously described C12 standards relate to networking and communication aspects. C12 contains additional specifications, for example:

- ANSI C12.01: "Code for Electricity Metering" defines some conditions for a set of tests performed on electricity metering equipment. C12.01 defines these tests and the associated acceptable performance criteria in terms of overload support, the effect of the variation of numerous parameters, the effect of heat, the effect of a magnetic field, and so on.
- C12.10: "Physical Aspects of Watthour Meters – Safety Standard". As mentioned in the NEMA abstract "This standard covers the physical aspects of both detachable and bottom-connected watthour meters and associated registers. These include ratings, internal wiring arrangements, pertinent dimensions, markings and other general specifications".
- C12.20: "Electricity Meters – 0.2 and 0.5 Accuracy Classes". This standard defines requirements for electricity meters in terms of voltage or frequency ranges, form designation, displays, and so on. Based on a list of 38 standardized tests, it also defines the acceptable performance for an electricity meter.

10.7   RFC 6142: C12.22 Transport Over an IP Network

RFC 6142 "ANSI C12.22, IEEE 1703, and MC12.22 Transport Over IP" provides a framework for transporting ANSI C12.22 data over an IP network. IEEE 1703 and MC12.22 are similar specifications in IEEE and Measurement Canada environments.

When ANSI C12.22 defines a layer 7 protocol and proposes to transport it over any underlying protocol, RFC 6142 proposes to restrict the scope to the transport of C12.22 messages by using TCP and UDP transports over an IP network. RFC 6142 more precisely focuses on the adaptation of Chapter 5 of the ANSI C12.22, which is related to the "C12.22 Node to C12.22 Network Segment Details".

All the C12.22 network elements considered in RFC 6142 are natively IP aware. In case of a C12.22 IP relay, RFC 6142 only deals with the IP interface and not with the other possibly non-IP aware interfaces that may used for message forwarding to C12.22 non-IP nodes.

RFC 6142 describes a C12.22 IP network segment in a general manner, without any intention to provide any guidelines on its size. A small LAN or a full C12.22 IP network are equally possible.

In order to convey the RFC 6142 C12.22 messages in a standardized way, port number 1153 was assigned by IANA (Internet Assigned Number Authority) for both TCP and UDP.

RFC 6142 specifies an encoding for the native IP address in order to standardize the use of IP addresses in the appropriate fields of ANSI C12.19 Tables. IPv4 and IPv6 are two possible options.

The support of IP multicast is required in all C12.22 hosts, relays and master relays and recommended in the C12.22 nodes in order to facilitate the reading of numerous C12.22 meters. In this case the meters have a common C12.22 multicast group ApTitle and can be reached by sending a single EPSEM read request. Two specific IPv4 and IPv6 multicast addresses have been assigned by IANA to a newly created "All C1222 Nodes" multicast group (224.0.2.4 for IPv4 address and FF0X::24 for IPv6). The use of a TTL (time to live) attribute in an IP packet header allows the propagation of C12.22 IP multicast messages to be limited.

C12.22 allows the use of two connection modes: a connection-oriented mode and a connection-less mode. RFC 6142 maps these modes to the use of TCP or UDP. For each type of C12.22 network elements, depending on their ability to support TCP or UDP and depending on their ability to be able to accept unsolicited new datagrams or connection requests, RFC 6142 defines a set of basic rules for correctly handling the application associations and the exchanges of UDP or TCP messages.

ANSI C12.22 contains its own security mechanism and does not mandate any transport layer security. RFC 6142 allows the use of a transport layer security mechanism as an enhancement to the C12.22 security feature.

10.8 REST-Based Interfaces to C12.19

Although this was not an explicit design principle, the design of C12 happens to be fully REST compliant. This is a lucky circumstance as modern smart-grid designs recommend the use of a REST style architecture. At present, no formal IP based REST interface for C12 has been proposed, however, a tentative interface for the ETSI TC M2M is presented in Chapter 14.

11

DLMS/COSEM

Jean-Marc Ballot,[*] and Olivier Hersent
*Alcatel-Lucent

11.1 DLMS Standardization

11.1.1 The DLMS UA

The Device Language Message Specification[1] user association was formed in 1997 by utilities and manufacturers to develop open standards for multiutility (all energy types) meter data exchange, for all application segments. As of 2010 it counts over 180 members, as well as multiple associate member organizations: ESMIG, M-Bus, Euridis.org, Selma, DVGW, PPCEM and the ZigBee Alliance. Over 140 meter types, from over 40 manufacturers, have been certified.

The DLMS UA maintains the specification, is the registration authority for IEC 62 056 (OBIS codes), performs technical support and training, and operates the conformance specification scheme. The DLMS UA is organized in two working groups:

- The Maintenance and Development WG, handling the development of the standard.
- The Final End Users and Developers WG, focused on use cases and gathering feedback from deployment and interop testing, led by French utility EDF.

11.1.2 DLMS/COSEM, the Colored Books

DLMS/COSEM separates the aspects of data modeling, data identification, messaging and transport:

---

[1] The initial name, from French Utility EDF, was "Distribution Line Message Specification".

The Internet of Things: Key Applications and Protocols, First Edition.
Olivier Hersent, David Boswarthick and Omar Elloumi.
© 2012 John Wiley & Sons, Ltd. Published 2012 by John Wiley & Sons, Ltd.

- COSEM, the companion specification for energy metering, specifies the data model, that is, the standard object interfaces, with their attributes and methods. It maintains a registry of object interfaces (OBIS data identification codes). xDLMS messaging is used to access COSEM objects attributes and methods.
- DLMS itself is an application layer protocol that defines abstract object-related services and protocols. Out of the original 22 services defined by DLMS, DLMS/COSEM uses only a subset of 4 messaging services, as well as a few extensions. This profile is named xDLMS.
- DLMS supports multiple transport layers: Twisted pair, power line, IP, and so on.

The DLMS specification is documented in 4 "books", which can be purchased on DLMS.com:

- The Blue book "COSEM – Identification System and Interface Classes" – DLMS UA 1000-1:2009, Ed. 9.0, 2009-02-09, specifies the data model (COSEM interface classes and OBIS codes for various energy types). The Blue book has been internationally standardized by IEC and CEN.
- The Green book "DLMS/COSEM – Architecture and Protocols" – DLMS UA 1000-2:2009, Ed. 7.0, 2009-12-22, specifies the protocols with DLMS on top, for the various media specific communication profiles. The Green book has been internationally standardized by IEC and CEN.
- The Yellow book specifies conformance test plans for COSEM object model.
- The White book is a glossary of terms.

Together, these books represent over 600 pages of specifications, and the specification is still rapidly expanding. The first implementation of DLMS/COSEM was deployed in 1999. In 2002 the specification was published as IEC and CEN standards. More recently, the standard was adopted by SM-CCG and OPENmeter consortium as the core standard for smart metering. DLMS/COSEM is also published as a standard in China and in India.

11.1.3 DLMS Standardization in IEC

The DLMS-UA work was co-opted by IEC TC13 (International Electrotechnical Commission, Technical Committee 13). The IEC TC13 is in charge of "electrical energy measurement, tariff and load control". In Europe, the CENELEC TC13 mirrors the IEC TC13. IEC TC 13 endorses the DLMS colored books in their IEC 62 056 series.

The contents of the "Green book" are reflected in the following IEC standard documents:

- **IEC 62 056-42:** Physical layer services and procedures for connection-oriented asynchronous data exchange;
- **IEC 62 056-46:** Data-link layer using HDLC protocol;

- **IEC 62 056-47:** COSEM transport layers for IPv4 networks;
- **IEC 62 056-53:** COSEM application layer.

The contents of the "Blue book" are reflected in the following standard documents:

- **IEC 62 056-61:** Object Identification System (OBIS).
- **IEC 62 056-62:** Interface classes.

The colored books are also reflected in CENELEC standards, for example, EN 13 757 part 1 for the Blue book COSEM interface object model.

11.2 The COSEM Data Model

COSEM is an object model for metering applications that is utility-type and communication-media independent. COSEM uses the client–server paradigm. In the COSEM model, the meter is the server. A COSEM server only models the elements of the meter that are visible externally. COSEM data structures are specified in ASN.1 syntax.

A COSEM server (a physical metering device) is modeled as a set of "logical devices", hosted in a single physical device. Each logical device models a subset of the functionalities of the physical meter. A logical device is implemented as an application process (AP).

A logical device is composed of a sct of COSEM interface objects. Interface objects model various functions of the meter and they are accessible from the client side through the communication interfaces of the meter. Each interface object is a collection of attributes and methods. The structure of objects that have common characteristics is described once for all in an interface class. The interface classes are specified in the DLMS Blue book.

These high-level data model principles are represented in Figure 11.1.

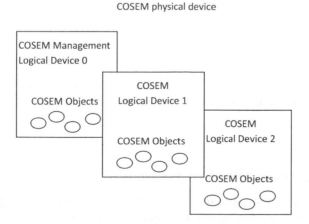

Figure 11.1 DMLS data model overview.

Each physical DLMS device shall provide one "management logical device" which contains the list of logical devices in the physical device ("service access point assignment") and may contain one or several other logical devices. Each logical device is identified by a unique COSEM logical device name object (LDN), an octet string of up to 16 octets starting with a 3-octet manufacturer identifier. Each logical device also contains an application association object, which contains in its object_list attribute the list of visible COSEM objects in the context of the application association session with a given client (the "association view").

Note: The DLMS and ZigBee high-level data models are very similar: DLMS Logical Devices are equivalent to ZigBee applications, DLMS interfaces objects are equivalent to ZigBee clusters. The DLMS Blue book is the equivalent of the ZigBee cluster library (ZCL). The fundamental difference is that ZigBee clusters behave as servers or clients, enabling peer to peer communication, while DLMS interface objects are only servers. Another difference is that the Blue book focuses on metering and basic meter related I/O control, while the ZCL scope extends beyond metering.

The object-oriented concept behind COSEM is that any real-world "thing" can be described by some attributes and methods. For the various applications (e.g., energy, billing, load profiles, instant power measurement, I/O control, access control), COSEM maps the "thing" to a COSEM object composed of:

– A set of attributes. Attributes have a meaning, a data type, a value range, and access rights.
– A set of methods (e.g., reset, start).

Similar objects make up an interface class (IC), specified using ASN.1 syntax.
    For instance, a meter can offer two registers:

– A register measuring the active energy $T1 = 1234$ kWh;
– A second register measuring the reactive energy $T2 = 0123$ kWh.

Both registers are very similar, and can be represented by a name, a value, a unit, and methods for example, reset. They can be modeled by the same interface class, the REGISTER class.

11.3   The Object Identification System (OBIS)

OBIS defines identification codes for commonly used data items in energy metering. The DLMS-UA Blue book and IEC 62 056-61 specify the overall structure of the identification system and the mapping of all data items to their identification codes.

Figure 11.2   Structure of an OBIS code.

All data items exposed by a meter are uniquely identified by an OBIS code. This is true for measurement data, but the scope of OBIS is wider than measurement data: OBIS codes also cover data used for configuration of metering equipment; or related to the meter status. OBIS codes are defined for all types of utility metering applications: electricity, water, gas and heating.

The concept of an OBIS code is based on a hierarchical structure composed of 6 different "value groups", from A to F (Figure 11.2 and Figure 11.3).

- Value group A is for the identification of the energy type. For example, electricity, hot or cold water, gas, heater, cooling are possible energy types.
- Value group B is used to distinguish between several possible inputs in a metering equipment. A data concentrator is a typical example of such equipment.
- Value group C is used for identifying the type of physical quantity, for example, voltage, volume, temperature, power. Value group C items clearly depend on the content of value group A.
- Value groups D and E are used for identifying additional data that can for example be the result of an internal processing by using a specific algorithm applied on data already identified from value groups A to C. It is also used for consortia- or country-specific applications.
- Value group F was originally planned for identifying historical data (billing periods) when needed. If there is no such data then value group E may be used for improving classification.

Obis code element	Range		Manufacturer extension codes
A	0–15	Abstract objects : 0	
		Energy type (**1**: electricity, **4** Heat, **5** Cooling, **6** Heat, **7** Gas, **8** Cold water, **9** Hot water)	
B	0–255	Channel	128–199
C	0–255	Type of physical quantity measured	128–199, 240
D	0–255	Processing, consortia or country specific	128–254
E	0–255	Classification	128–254
F	0–255	Historical	128–254

Figure 11.3   OBIS code value groups.

A part of the range of values in value groups B to F are reserved for manufacturer-specific data.

OBIS codes can be represented by 6 integers in dotted A.B.C.D.E.F format, for example, 1.0.1.8.0.255.

Thousands of OBIS codes have been defined, the complete list is available on the DLMS user association web site (DLMS.com). Interface classes are versioned, and can be extended over time.

In the case of very simple devices, the logical name (LN) referencing method using the OBIS system is not used. A simpler system, using a 13-bit integer for referencing any attribute of a COSEM interface object is used. This simpler system is called short name (SN) referencing. This is useful for ensuring compatibility with the older versions of DLMS.

11.4   The DLMS/COSEM Interface Classes

COSEM defines standard objects, defined by their interface classes (IC), for data storage, access control and management, time and event bound control. Interfaces classes are specified in the DLMS-UA Blue book and in the IEC 62 056-62.

Each interface object is a collection of attributes and methods:

- Attributes represent the characteristics of the object. The first attribute, mandatory, is the "logical name" and its value is an OBIS code or a short name that identifies the measurement category applying to the object instance. For instance, a "register" object with logical name [1 1 1 8 0 255] measures electric total positive active energy, while a "register" object with logical name [1 1 3 8 0 255] measures electric total positive reactive energy. Each interface class definition also allocates an index to each attribute. Each attribute is uniquely identified:
  - by the class ID and "logical name" of the object instance to which it belongs, and its index within this instance (LN referencing)

  or
  - by a short 13 bit-integer (SN referencing), for simple devices. Some SN values are reserved for special objects, for example, 0xFA00 for the Association SN.
- **Methods:** in the object-oriented model of DLMS, external entities can act on the object only through defined methods, for example, for accessing attribute values. For instance, the "reset" method, on a register interface class, sets the current consumption value to the default value. More complex methods are defined, for instance methods that trigger authentication procedures. Within an object instance, methods are identified by their index (LN referencing) or by a short integer (SN referencing).

The set of interface classes represents a tool box that a manufacturer can use when building a meter product, and facilitating interoperability. The model can be extended, new objects only need to be added to the OBIS registry and defined using the appropriate ASN.1

description. Some OBIS code ranges are reserved, for example, for national extensions (specific attributes, interface classes) by using the E164 country code in field C of the interface OBIS code.

Manufacturers may decide not to implement standard interface classes for all objects and use the DLMS manufacturer extension mechanisms. However, when a standard interface class is used, it must be implemented in conformance with the DLMS Blue book.

11.4.1   Data-Storage ICs

- **Register (class ID 1):** this object contains a value, and an enumerated pointer to a unit.
- **Extended register (class ID 3):** this object extends the register by providing a time stamp.
- **Demand register (class ID 5):** extends the register object by storing the current value, as well as maximum and minimum values.
- **Register activation (class ID 6):** this object specifies at which periods of the day which register is activated.
- **Profile generic (class ID 7):** this is a generic "spreadsheet-like" object.
- **Utility table (class ID 26):** this IC encapsulates ANSI C12.19 table data. Each "table" is represented by an instance of this IC, identified by its logical name. The IC attributes are the ANSI Table-Id, the length of the table, and a buffer containing the table data.
- **Register table (class ID 61):** a simpler version of the profile generic object, which can be used to store multiple similar values.
- **Status mapping (class ID 63):** while status codes can hardly be standardized, this table maps custom-status codes to utility-specified values.

11.4.2   Association ICs

These objects are specified as gatekeepers to other objects:

- **Association SN (class ID 12):** list of SN references to objects of a given logical device that are accessible in a given association context with a COSEM client. This object may be present multiple times if a logical device supports multiple application associations.
- **Association LN (class ID 15):** same as above using LN referencing.
- **SAP Assignment (class ID 17):** the service access point assignment object contains the list of logical devices within a physical device and their respective service access points.
- **Image transfer (class ID 18):** this object is used to manage the upload of software images.
- **Security setup (class ID 64):** contains information on security policies within a particular application association, and methods to set up security keys.

11.4.3   Time- and Event-Bound ICs

- **Clock class (class ID 8):** the clock object, including timezone and daylight saving data.
- **Script table (class ID 9):** scripts that can be used for the activation of tariffs, upload of a new firmware, and so on ... Scripts are a sequence of method invokes or attribute modifications.
- **Schedule object (class ID 10):** the "to do list" object, specifying time- or date-driven activities.
- **Special days table (class ID 11):** list of special days for use with the schedule object or the activity calendar.
- **Activity calendar (class ID 20):** defines a calendar-based schedule of actions.
- **Register monitor (class ID 21):** can be used to configure the monitoring of values of several registers and, if certain triggers are met, to execute action scripts.
- **Single action schedule (class ID 22):** for example, execute firmware.
- **Disconnect control (class ID 70):** manages a disconnect unit of the meter, for example, a contactor.
- **Limiter (class ID 71):** triggers an action script when the value attribute of a monitored object crosses a threshold for a certain amount of time.

11.4.4   Communication Setup Channel Objects

Multiple objects have been defined to manage the physical layer parameters and communication setup over these physical layers, for instance:

- IEC local port for IEC 62 056-21 ports;
- IEC HDLC setup;
- TCP-UDP setup;
- IPv4 setup;
- IPv6 setup;
- M-Bus slave.
- M-Bus client (meter acts as master), enable mapping of M-Bus data identifiers (data information block, variable information block) to M-Bus value objects of "extended register" interface class objects.
- M-Bus master port setup, to set EN 13 757-2 interfaces.

11.5   Accessing COSEM Interface Objects

11.5.1   The Application Association Concept

In order to allow the client party to access COSEM interface objects in the server, the DLMS-UA defined the concept of "application association". This application association is an application-level connection. It is established between a Client AP (application process) and a server AP (one of the logical devices that are modeled in the metering equipment). There is only one Association per logical device. The client AP always initiates the establishment of the association. For very simple devices, one-way communicating devices, and for multicasting and broadcasting pre-established associations are also allowed.

During the association establishment, some contextual data is exchanged and the authentication mechanisms are selected.

After the association establishment, the client AP and the server AP can exchange application data: some of the COSEM interface objects in the server (i.e. one of the logical devices of the metering equipment) become accessible for the client AP. Several data communications services are specified in order to exchange data. Once data exchanges are finished, the association has to be released.

The association establishment is performed by using some basic services of the COSEM application layer that is presented in the next section.

11.5.2   The DLMS/COSEM Communication Framework

The DLMS/COSEM protocol stack contains a metering application, the COSEM application layer and COSEM transport layers. The COSEM application layer is unique for any type of transport layer. Data are exchanged between a server AP and a client AP by using communication profiles (one in the server, one in the client). DLMS-UA defined several communication profiles (implemented in the COSEM transport layer) in the DLMS Green book (IPv4, HDLC, PLC, M-Bus, . . .).

Note: The DLMS version used in DLMS/COSEM is an extension of the original DLMS specified in IEC 61 334-4-41. This extended version is referred to as xDLMS. However, in the text, we continue to use DLMS.

For a better readability, only the IPv4 transport layer is represented in Figure 11.4.

The COSEM application layer provides a set of services in order to access to the application interface objects and methods. COSEM application layer services are split into 3 categories:

− application association establishment and release;
− data transfer;
− layer management (for local management, then out of scope of DLMS specifications).

Due to the existence of two different referencing methods (LN and SN) for accessing the meter objects, the COSEM application layer in the client side contains two different

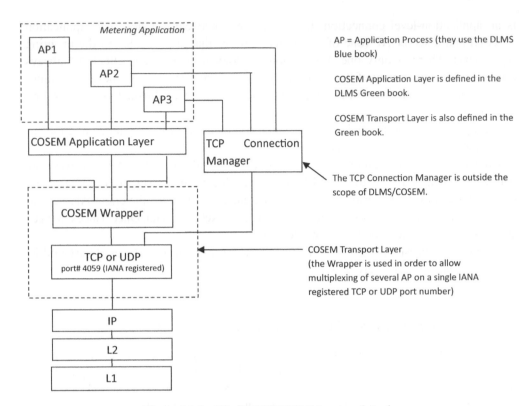

Figure 11.4 The DLMS/COSEM protocol stack.

sets of services: one for the logical referencing method, the other one for the short name referencing method.

The DLMS application protocol is connection oriented: in the previous section we explained that an application association has to be established between a server AP and a client AP before any communication with COSEM objects can occur. The set of services in charge of application association handling is composed of three services:

- COSEM-OPEN.request;
- COSEM-RELEASE.request;
- COSEM-ABORT.request.

The principle is the following:

- COSEM-OPEN.request sets up an application association. During the association establishment, a specific COSEM interface object is created: the "association" object. Among several attributes, this association object contains the list of all visible COSEM interface objects in the context of this association: after Association establishment, the client application process can read the list of visible interface objects, and perform some operations on these objects.

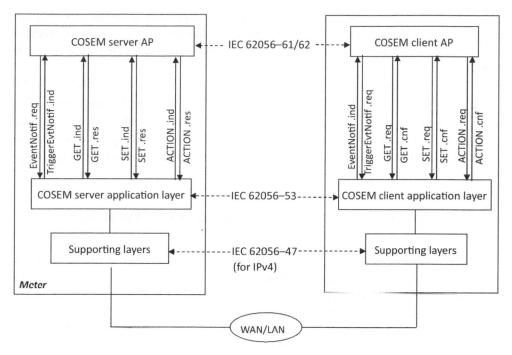

Figure 11.5 DLMS data communication services (LN referencing case).

- Application data exchange takes place using DLMS data communication services (refer to the next section for details).
- COSEM-RELEASE.request releases the application association.

for pre-established AAs, OPEN/RELEASE/ABORT requests are not used.

11.5.3 The Data Communication Services of COSEM Application Layer

The data communications services applicable to LN referencing are summarized on the Figure 11.5.

The set of services is: GET, SET, ACTION and EventNotif.

- The GET service is invoked by the client AP to request the value(s) of one or all attributes of one or more COSEM interface objects from the server AP. For example, the GET service is used for reading the value of an electricity counter. In this case, the class_id is 3 (for register class), and the attribute targeted by the GET service is the "value" attribute.
- The SET service is invoked by the client AP to request the remote server AP to set the value of one or more attributes of a COSEM interface object. For example, the SET service may be used for changing the electricity tariff for a specific period of time.

- The ACTION service is used by the client AP to remotely invoke one or more methods of one or more COSEM interface objects in the remote server AP. The server AP executes the requested ACTION. The reset of a register is a typical example of an invokable method.
- The EventNotification Service: this service is used in order to enable the server AP to send an unsolicited notification of the occurrence of an event to the remote client AP. This notification contains the value of a COSEM interface object attribute. It is an exception to the client–server paradigm. The client AP may explicitly solicit sending an EventNotification message by calling the Trigger_EventNotification_sending service primitive.

In the case of SN referencing, the list of data communication services is different.

- The read service is used to read the value of one or more attributes or to invoke one or more methods of COSEM interface objects.
- The write service is used to write the value of one or more attributes or to invoke one or more methods of COSEM interface objects.
- The UnconfirmedWrite service is used to write the value of one or more attributes or to invoke one or more methods of COSEM interface objects. It is an unconfirmed service.
- The InformationReport service: upon the occurrence of a specific event, the server can inform the client party of the value of one or more COSEM interface object attributes. It is an exception to the client–server paradigm.

The parameters for read/write must include:

- the physical layer MAC address of the meter (this is used by lower layers to establish communication with the meter or concentrator);
- an InvokeID also encoding the priority of the message;
- the interface class OBIS code, for example, REGISTER class;
- the interface class instance (e.g., multiple REGISTER classes may exist on a meter);
- the identifier of the attribute (for get and set), or the identifier of the method (for action).

Referencing may also use short-name mapping of logical name using the interface class mapping table.
    The responses include:

- the destination physical layer MAC address;
- an InvokeID also encoding the priority of the message;
- the response data.

Data formats are described in the OBIS profile for the relevant object class. For instance, for the active energy register, attribute 2 is used to store the register using long unsigned encoding, while attribute 3 is an enumerated value mapped to the physical unit.

The lower layers encode the DLMS messaging primitives to PDUs, using A-XDR encoding, a specific version of ASN.1 BER optimized for COSEM data types specified in IEC 61 334-6.

11.6   End-to-End Security in the DLMS/COSEM Approach

DLMS/COSEM provides security features in two different domains:

- Access control security: controls the server data that a given client may access using role-based access rules.
- Security for data transport: provides security during the transport of data from a DLMS/COSEM end-point to another DLMS/COSEM endpoint.

11.6.1   Access Control Security

Access control security is provided as part of the application association establishment procedure.

In order to be able to access server side data, the client has to be authenticated. This is performed during the association establishment. Depending on the capabilities of the meter, the level of the security for the data access is negotiated. DLMS/COSEM provides three different levels of data access security:

- Lowest-level security: in this case there is no security at all. Peer authentication is not needed. This level allows direct access to the data contained in the server.
- Low-level security (LLS).
- High-level security (HLS).

In the LLS security model, the security is ensured via a username/password scheme. The goal is not the authentication of the server. Only the client is authenticated by providing a secret (generally a password) during the application association establishment procedure. The server checks whether the password is correct then the association is considered as established.

The association interface class provides a way to access the password in the server by using the "change_secret" method.

In the HLS security model, a mutual authentication is a prerequisite for application association establishment. Different HLS_Authentication_Mechanisms may be negotiated during the application association establishment (e.g., with different methods for generating a digest, based on MD5, SHA-1, ...).

Once the client is authenticated, the list of objects that may be accessed is determined by the server and presented in the AA object_list attribute. This doorkeeper function controls access to associations, registers, profiles, clocks, and so on, using access tables is according to the requester role determined by its identity.

11.6.2   Data-Transport Security

This part of the security scheme provides cryptographic data protection. Ciphering and deciphering is performed by the COSEM application layer on a per-message basis. In order to decide whether ciphering protection is needed, the COSEM AL uses information contained in the security context that was negotiated during the application association establishment. The security context is contained in a security setup object associated to the application association and specifies:

- The level of security to be applied to messages:
  - no security;
  - all messages have to be authenticated;
  - all messages have to be encrypted;
  - all messages have to be both authenticated and encrypted.
- The security algorithm to be used: currently the DLMS specifications contain only one security suite, the Galois/counter mode (GCM) with AES-128 symmetric encryption algorithm. Some additional security suites may be added in the future.
- The different security materials and credentials: among them, the master key, the ciphering keys, the authentication keys, the initialization vectors, . . .
- Security setup object linked to association object, specifying which security services to apply (e.g., encryption).

All meters must have a master key that is pre-established (and communicated via database transfer to the meter controller).

Part Four

The Next Generation: IP-Based Protocols

12

6LoWPAN and RPL

12.1 Overview

Traditionally, battery-powered networks or low-bitrate networks, such as most fieldbus networks or 802.15.4 (see Chapter 1 for details) were considered incapable of running IP. In the home and industrial automation networks world, the situation compares to the situation of corporate LANs in the 1980s: "should I run Token-Ring, ATM or IPX/SPX?" translates to "should I run ZigBee, LON or KNX?"

IP, with its concept of layer 3 routing and internetwork technology, has made those debates about incompatible networks obsolete: the vast majority of LANs and WANs today run IP, and many people can hardly remember which layer 2 technology their IP networks are running on. Almost any layer 2 technology can be used and will simply extend the IP internetwork.

The same transition to IP is now happening in the home and industrial automation worlds. 6LoWPAN and RPL have made this possible.

12.2   What is 6LoWPAN? 6LoWPAN and RPL Standardization

The Internet Engineering Task Force (IETF) 6LoWPAN Working Group was formed in 2004 to design an adaptation layer for IPv6 when running over 802.15.4 low-power and lossy networks (LowPAN or LLN). The work included a detailed review of requirements, which were released in 2007 (RFC 4919).

In practice, however, the 6LoWPAN is not restricted to radio links, and the technology can be extended to run over other media, for instance it has been extended to run over low-power CPL (www.watteco.com) or G3 OFDM CPL. IPv6 is also being adapted to other physical layers, independently of 6LoWPAN, for example, for Home-Plug CPL. Many fieldbus vendors are now considering an IPv6 adaptation layer for their products.

The Internet of Things: Key Applications and Protocols, First Edition.
Olivier Hersent, David Boswarthick and Omar Elloumi.
© 2012 John Wiley & Sons, Ltd. Published 2012 by John Wiley & Sons, Ltd.

802.15.4 and most low-power transmission technologies must rely on mesh networking to create large networks. Two techniques may be used:

- "Mesh under": the link layer (layer 2) supporting the IP network takes care of mesh networking and packet forwarding, and the IP layer sees a large subnet. An example of such a mesh under protocol is GeoNET, currently under development to support car to car transmission as part of the ETSI intelligent transport system (ITS) technical committee (http://www.geonet-project.eu/, http://www.etsi.org/website/Technologies/ IntelligentTransportSystems.aspx). Mesh under is also used in the large 6LoWPAN backbone of the smart metering project of France DSO ERDF.
  Obviously, mesh under works only within the context of a single link-layer technology.
- "Route over": IP level (layer 3) mesh routing. If multiple underlying networking technologies need to be used simultaneously (e.g., wireless 802.15.4 and CPL), or when the underlying networking technology supports only point to point or local broadcast link layer communication capabilities, then IP level mesh routing becomes necessary to form the internetwork.

The IETF Routing Over Low-power and Lossy networks (ROLL) Working Group was formed in 2008 to create such an IP level routing protocol adapted to the requirements of mesh networking for the Internet of Things: the first version of RPL was finalized in April 2011 (at the time of writing the RFC was not yet allocated). See Section 12.4 for more details.

At present the reference documents for 6LoWPAN are:

- IETF RFC4919: "IPv6 over Low-Power Wireless Personal Area Networks (6LoW-PANs): Overview, Assumptions, Problem Statement, and Goals".
- IETF RFC4944: "Transmission of IPv6 Packets over IEEE 802.15.4 Networks".
- Internet draft ID-6LoWPAN-HC defines HC (header compression), which will replace the header compression mechanism defined in RFC4944 (now "NOT RECOMMENDED"). This document is in "RFC queue" status at the time of writing.
- draft-ietf-6LoWPAN-nd-16, which updates the original IPv6 neighbor discovery mechanism for use in LowPANs.

Some existing networks use alternative pre-RPL routing methods, such as IETF draft-daniel-6LoWPAN-load-adhoc-routing-03: "6LoWPAN *ad hoc* on-demand distance vector routing (LOAD)", used in the French AMI initiative of ERDF.

12.3   Overview of the 6LoWPAN Adaptation Layer

6LoWPAN is designed to work on top of 802.15.4 networks. The optional hop by hop acknowledgment feature of 802.15.4 is used, but the macMaxFrameRetries should be set to a relatively low value (e.g., the default of 3) in order to make sure the 802.15.4 layer will not continue to retry when IP and application-level retransmission mechanisms trigger.

6LoWPAN needs to solve 4 issues:

- Header compression: on battery-powered networks, long packet headers is synonymous with energy waste. Native IPv6, with its 40-byte header, was probably one of the worst possible candidates for such networks: without compression, the payload of a single IPv6 UDP packet transmitted over a 802.15.4 link layer would not be able to exceed 53 bytes! In the most favorable case, the LowPAN and UDP compressed headers require just 6 bytes.
- Packet fragmentation and reassembly: low-power networks usually provide small MTUs, because transmission uses energy, and transmission time is proportional to the packet size. Also, small packets are less subject to packet loss that may occur over lossy networks such as 802.15.4. For instance, on 802.15.4 networks, the frame size is only 127 bytes, and the MAC level overheads (addressing fields, FCS, security headers, see Chapter 1 for more details) may leave as little as 81 bytes for IP. IPV6 normally requires a MTU of 1280 bytes!
- Adaptation of IPv6 neighbor discovery defined in RFC4861 and 4862.
- Support for "mesh under" layer 2 forwarding.

One of the issues of 802.15.4 is that it forgot to define a field to identify the "next higher protocol" (e.g., the equivalent of the Ethernet "Ethertype" field). Therefore, there is no reliable mechanism to share a given 802.15.4 PAN among multiple L3 protocols, like 6LoWPAN and ZigBee 1.0.

6LoWPAN currently defines several headers, which appear in the following order when present:

- The mesh addressing header;
- Hop by hop processing header, which encode hop-by-hop options such as BC0 broadcast sequence number;
- Destination processing: for example, the fragment header;
- Payload transport: for example, the IPv6 and UDP compression headers.

The first byte of each header, called the dispatch byte, identifies the nature of the header (Figure 12.1). A large subset of the dispatch byte space is currently reserved, leaving some room for future 6LoWPAN extensions or future coexistence with other protocols that would use the same dispatch byte.

12.3.1 Mesh Addressing Header

Currently, no "mesh-under" protocol is defined for 802.15.4, so this header is only a facility provided to make it possible in the future. When 802.15.4 mesh-under routing is

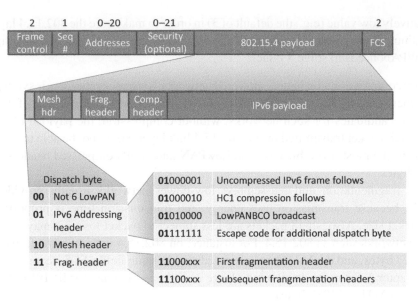

Figure 12.1 6LoWPAN header stacking, the dispatch byte.

enabled, the 802.15.4 MAC frame contains the source and destination addresses for each hop, therefore a container is needed for the original and final 802.15.4 addresses. The mesh addressing header provides such a container, and also contains a "HopLeft" counter that should be decremented by each layer2 hop.

12.3.2   Fragment Header

The fragment header for the first fragment specifies the full (reassembled) packet size, and uses a datagram tag common for all fragments of this IP packet, which will be used by the receiver, together with the sender and destination MAC addresses, to identify fragments belonging to the same packets. Subsequent fragments also specify the offset of the fragment in the full IP packet, in multiples of 8 bytes (see Figure 12.2).

12.3.3   IPv6 Compression Header

12.3.3.1   Forming an IPv6 Unicast Address from the 802.15.4 EUI64 or 16-bit Short Address

The 802.15.4 EUI64 is composed of a 24-bit OUI (organizationally unique identifier) and a 40-bit extension identifier chosen by the manufacturer). The OUI has two reserved bits in its first octet: the least significant bit is reserved to define a space for multicast addresses, and the second least significant bit (L) is used to distinguish locally assigned addresses from universal addresses formed as OUI+extension.

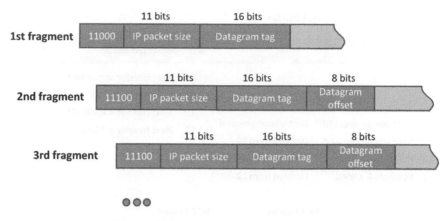

Figure 12.2   6LoWPAN fragment header.

In order to form the interface Id (IID), RFC4291 chooses to invert the L bit of the EUI64 and calls it the U bit (universal address bit), so that locally assigned addresses will have all-zero prefixes and be more compact and easier to remember.

6LoWPAN mandates that IPv6 local addresses will be derived from the EUI64 addresses using the following convention: 64 bit prefix + Ubit formatted EUI64. Prefix FE80:: (1111111010 followed by 54 zeroes) is used for link local addresses, prefixes beginning with 001 are global prefixes for unicast addresses.

The IPv6 address can also be derived from the 16-bit 802.15.4 short address, by concatenating it with the PAN identifier, or with 16 zeroes if the PAN ID is unknown (RFC 4944 Section 6). 6LoWPAN requires bit 6 (Ubit) of the PAN identifier to be zero as it is not a universal address. Short addresses beginning with a 0 bit are reserved for unicast addresses, and short addresses starting with 100 are reserved for multicast addresses. Other values are reserved.

12.3.3.2   HC1-HC2 Compression (Now "Not Recommended" and Replaced by HC)

The original 6LoWPAN standard (RFC4944) defined a simple stateless compression mechanism compatible with the capabilities of resource constrained nodes, which exploits the redundancies between the MAC layer and the IPv6 layer, and encodes most likely values of variable fields in a more compact format. The "v6" version field is elided.

Figure 12.3 shows the conventions used by HC1 and HC2. The "C" flag indicates that the flow label and traffic class are all zeroes and elided.

When the payload is UDP (NH=01), then HC2 compression can be used to reduce the size of UDP ports (only 4 bits are sent inline if the ports are in the 61 616–61 631 range), and to omit the redundant UDP size field.

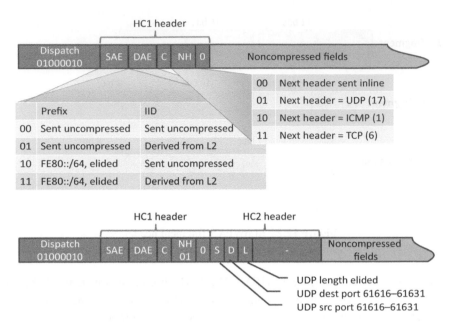

Figure 12.3   HC1, HC2 compression.

The noncompressed IP fields are sent starting with the hop counter, then in the order of the HC1 header elements. Uncompressed UDP fields are sent in the order of standard UDP fields.

12.4   Context-Based Compression: IPHC

HC1 compression works well when using link-local addresses, but any communication with IPv6 nodes located outside the local network will require globally routable IPv6 addresses, which are not compressed with HC1. Internet draft ID-6lowpan-hc defines a new header for IPHC compression with simple support for shared context information between sender and receiver.

IPHC "steals" 5 bits out of the reserved dispatch value field for its own 13-bit base header, outlined in Figure 12.4.

The new IPHC header format adds an ability to selectively compress IPv6 flow labels (RFC2460 and RFC3697), 6-bit differentiated services code points (DSCP, RFC2474 and RFC3260) and explicit congestion notification (ECN, RFC 3168). The address compression takes into account prefixes indexed by the optional context IDs (SCI and DCI).

If the N bit is set, the IPHC header and uncompressed IPv6 fields are followed by a LOWPAN_NHC header, otherwise the IPv6 next headers are transmitted inline.

The format of the LOWPAN_NHC header is illustrated in Figure 12.5.

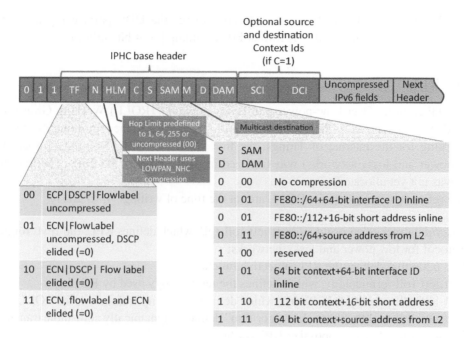

Figure 12.4 The IPHC 6LoWPAN header structure.

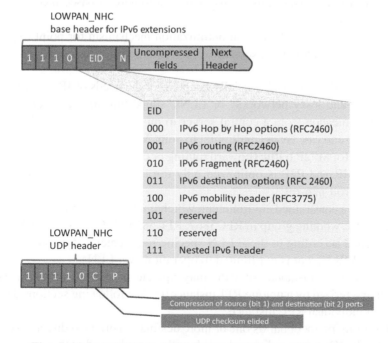

Figure 12.5 LOWPAN_NHC format for UDP and IPv6 options.

The LOWPAN_UDP header optionally compresses the UDP ports using the same approach as HC2 (compressed port format F0xx sending just 4 bits inline).

12.5 RPL

The IETF Routing Over Low-power and Lossy networks (ROLL) Working Group was formed in 2008 to create an IP level routing protocol adapted to the requirements of mesh networking for the Internet of Things: the first version of RPL (Routing Protocol for Low-power and lossy networks) was finalized in April 2011 (at the time of writing the RFC was not yet allocated).

The reference documents for ROLL are (at the time of writing, April 2011):

- https://datatracker.ietf.org/doc/draft-ietf-roll-rpl/, which defines RPL, the IPv6 routing protocol for low power and lossy networks;
- RFC 6206 that defines the RPL objective function 0;
- draft-ietf-roll-terminology, which defines the terminology used by ROLL;
- draft-ietf-roll-security-framework, which defines a security framework for ROLL;
- draft-ietf-roll-trickle, "the trickle algorithm" defines a dynamically adjustable transmission window scheme to optimize RPL traffic;
- draft-ietf-roll-routing-metrics for the computation of metrics;
- draft-ietf-roll-minrank-hysteresis-of that defines a hysteresis-based mechanism to prevent topology oscillations;
- draft-ietf-roll-p2p-rpl that defines an optimization mechanism for point to point communication (e.g., in automation scenarios for sensor/actuator messages flows).

RPL specifies a routing protocol specially adapted for the needs of IPv6 communication over "low-power and lossy networks" or LLNs, supporting peer to peer traffic (point to point), communication from a central server to multiple nodes on the LLN(point to multipoint P2MP) and *vice versa* (multipoint to point MP2P). The base RPL specification is optimized only for MP2P traffic (upward routing or convergecast used, e.g., in metering networks) or P2MP, and P2P is optimized only through use of additional mechanisms such as draft-ietf-roll-p2p-rpl.

Such LLNs are a constrained environment, which imply specific requirements explored by the IETF ROLL working group in RFC5867, RFC5826, RFC5673, and RFC5548. RPL has been designed according to these LLN specific requirements (typically on networks supporting 6LoWPAN), but is not limited to operation over LLNs.

Multiple concurrent instances of RPL may operate in a given network, each RPL instance is characterized by a unique RPLinstanceID. The following sections describe the behavior of an individual RPL instance.

The RPL routing protocol builds one or more destination oriented direct acyclic graphs (DODAG). Each DODAG is a directed graph with no cycles and with a single root node (see Figure 12.6). The graph is built according to optimization objectives specified by an

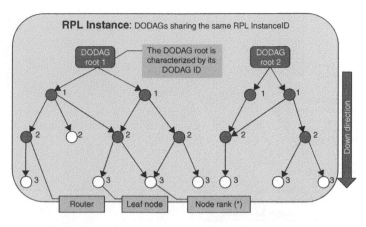

(*) : the node Rankstrictly increases in the Down direction. The exact way Rank is computed
depends on the DAG's Objective Function (OF), and is valid for a specific DODAG version

Figure 12.6   RPL builds a destination-oriented direct acyclic graph (DODAG).

objective function (OF, defined by the OCP field of a DIO DODAG configuration option).
The objective function is not specified by RPL itself, but in other companion documents
according to domain-specific requirements: for the available network metrics, the OF
computes the "rank" measuring the "distance" between the node and the DODAG root
and also defines the parent node selection policy, for instance an objective function could
seek to minimize the expected packet delay, while another might want to avoid routing
through any battery-operated node (see [I-D.ietf-roll-routing-metrics]).

RPL requires bidirectional links. Bidirectional connectivity must be verified before
accepting a router as a parent, for example, by using IPv6 neighbor unreachability detection

ICMPv6 Type=155	Code	Checksum		
Security (secure RPL msgs only)		0x00	DODAG Information Sollicitation	
		0x01	DODAG Information Object	
		0x02	Destination Advertisement Object	
Base		0x03	Destination Advertisement Object Ack	
		0x80	Secure DODAG Information Sollicitation	
		0x81	Secure DODAG Information Object	
		0x82	Secure Destination Advertisement Object	
options		0x83	Secure Destination Advertisement Object Ack	
		0x8A	ConsistencyCheck	

Figure 12.7   Structure of ICMPv6 RPL control message.

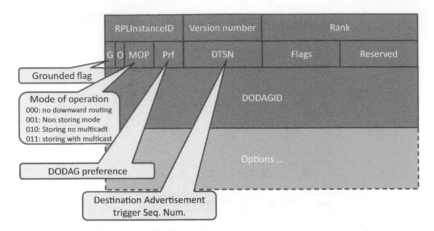

Figure 12.8 RPL DIO base object (followed by options).

(NUD), bidirectional forwarding detection (RFC5881) and hints from lower layers via layer 2 triggers like RFC5184.

12.5.1   RPL Control Messages

RPL routers need to exchange information in order to build the DODAG and populate routing tables. RPL defines a new ICMPv6 (RFC 4443) message, type 155, for this purpose.

RPL defines the following base objects:

- The DODAG information solicitation (DIS) message;
- The DODAG information object (DIO), see Figure 12.8;
- The destination advertisement object (DAO);
- The DAO Ack object;
- The consistency check (CC) object, which is used to check secure message counters and to carry RPL challenges and responses, and is always carried in a secure RPL message.

12.5.2   Construction of the DODAG and Upward Routes

The DODAG information object (DIO) is used to build the DODAG: it carries general DODAG configuration parameters and information that allows listening RPL routers to select a set of DODAG parents. Several type-length-value encoded options in the same RPL control message may specify:

- The address of the sending RPL router, and prefixes that may be used for IPv6 stateless autoconfiguration (0×08 prefix information option, or PIO). The PIO contains the same

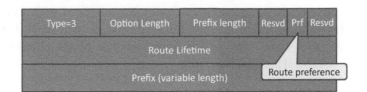

Figure 12.9 RPL Route Information option.

fields as the IPv6 neighbor discovery prefix information option defined in RFC4861, RFC4862 and RFC3775. A 1-bit "L flag" indicates that addresses derived from the prefix can be considered "on-link", a 1-bit "A flag" indicates that the prefix can be used for stateless address autoconfiguration.

- Metrics allowing estimation of the cost to reach destinations starting with each prefix (0×02 metric container option, formatted as specified in ID.IETF-roll-routing-metrics),
- One or more prefixes that are reachable by the advertising node (0×03 routing information option, illustrated in Figure 12.9 and containing the same fields as the IPv6 neighbor discovery route information option defined in RFC4191).
- Additional DODAG configuration information (0×04 DODAG information option) such as the values of MaxRankIncrease and MinHopRankIncrease used to constrain the rank a node can advertise when reattaching to a DODAG, or the default lifetime of all RPL routes.

RPL nodes send DIOs periodically via link-local multicasts, and joining nodes may request DIOs from their neighbors by multicasting ICMPv6 control messages containing a DODAG information solicitation Object (DIS). DIO parameters are explained in Figure 12.8, the DTSN is an 8-bit unsigned integer number set by the issuer of the message. In the storing mode of operation, incrementing the DTSN is a way to request updated DAO messages from child nodes.

Each DODAG, identified by a unique RPLInstanceID and DODAGID, is built incrementally from the root to leaf nodes:

- RPL nodes, starting by the DODAG root, advertise their presence, affiliation with a DODAG, routing cost, and related metrics by sending link-local multicast DIO messages to the all-RPL-nodes address. The DODAG root advertises predefined rank ROOT_RANK (=MinHopRankIncrease), and also specifies if it is "grounded", that is, if it can reach the set of destinations specified by the local DODAG policy (the "goal"). A DODAG is said to be floating if it cannot satisfy the goal.
- Nodes use the received DIO information to join a new DODAG and select their parents in the DODAG, or to maintain their affiliation to an existing DODAG. Nodes select parents according to the policy specified by the objective function and the rank of their neighbors as advertised by DIO messages. For the determination of parent

relationships, the ranks of potential parent nodes are compared with a granularity of MinHopRankIncrease (specified in the DIO messages), so that parent1 and parent2 will be considered of equal rank if floor(rank(parent1)/ MinHopRankIncrease) = floor(rank(parent2)/MinHopRankIncrease).

A first set of nodes will attach to the DODAG root and start to advertise DIO messages with the corresponding RPL instance and DODAG ID, expanding the reach of the DODAG. As new nodes will start to hear the RPL instance DIO messages and attach to it, the DODAG reach expands further until it reaches all nodes willing to attach to this RPL instance.

Nodes provision upward routing table entries according to their local policies (e.g., least cost), for the destinations specified by the DIO message, setting one or more DODAG parents as the next hop.

Each DIO announcement is attached to a specific DODAG version, therefore if the root decides to change the DODAG version it triggers a complete recalculation of the DODAG topology: this is a global DODAG repair.

A node may poison previously announced routes by advertising a special rank value of INFINITE_RANK (=0xFFFF). Note that if the destination cannot be reached temporarily, the node should rely on the local repair procedure and not poison the routes. The node may also decide to create a floating DAG. Poisoning a route implies that all sub-DAGs will also have infinite rank and therefore breaks the DAG topology.

12.6 Downward Routes, Multicast Membership

RPL uses destination advertisement object (DAO, Figure 12.10) messages to establish downward routes and to indicate multicast group membership. DAO messages are not mandatory and are required only by RPL instances that provide support for point to multipoint or peer to peer traffic. RPL control messages carrying a DAO object may also transport a list prefixes announced as reachable RPL targets or multicast groups (0×05 RPL target option), opaque RPL target descriptors (0×09 RPL target descriptor option),

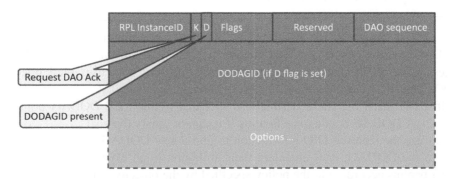

Figure 12.10 DAO object format.

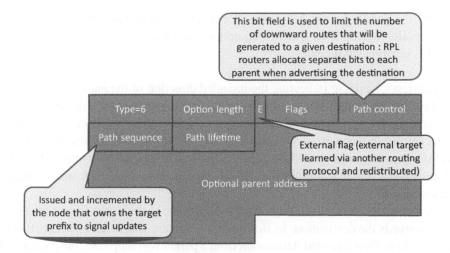

Figure 12.11   RPL transit information option.

and transit information (0×06 transit information option, Figure 12.11) which is used to indicate attributes for a path to one or more destinations, for instance its lifetime (a lifetime of 0×00000000 indicates loss of reachability to a target).

RPL supports two models for downward routing, each RPL instance supporting downward traffic selects one of the two models:

- In the storing model, RPL routing nodes are stateful. DAO messages, including the prefixes and addresses reachable by the sending node, are sent to the parents. Parents store the preferred downward routes and propagate aggregated DAOs upward.
- In the nonstoring model, all downward traffic includes a source routing header specifying each hop along the path, and intermediary routers do not store any routing information. Nodes send unicast DAO messages to the DODAG root, which include prefixes and addresses directly reachable by the node, and the node parents (in a transit information option as illustrated in Figure 12.11). The set of DAO messages enables the DODAG root to calculate an optimal hop by hop source routing path for each advertised destination. This mode has important implications: messages will be much longer (include source routing information), and P2P traffic is always routed to the DAG root.

12.7   Packet Routing

IP packets injected in an RPL network must have a RPL header that specifies the RPL instance, except when strict source routing is used.

If some leaf nodes send IP packets without such an RPL header, the first RPL router is required to add it, and select a default RPL instance. The RPL specification in itself does not specify the header format, but points to the ID-ietf-6man-rpl-option that places the RPL information into an IPv6 hop-by-hop option header.

The RPL information includes:

- "O": the down 1-bit flag indicating the intended direction of the packet.
- "R": the Rank-Error 1-bit flag signaling that a mismatch has occurred during forwarding between the rank relationship of the sender and receiver, and the effective direction of the packet. Such inconsistency, which can happen during the construction of a new DODAG version of a given RPL instance, is allowed to happen only once. RPL routers will absorb packets that are already "R" flagged in case such rank inconsistency is detected again.
- "F": the forwarding error 1-bit flag which indicates the node cannot forward the packet further towards the destination. In storing mode, routers that receive packets that they cannot route to their intended destination from a parent will loop back the packet to this parent with the F flag set: this allows the parent to remove the erroneous DAO routing entry from its routing table ("DAO insconsistency detection and recovery").
- The 8-bit RPLInstanceID.
- The 16-bit SenderRank, which must be set to zero by the packet source and then is set to the rank value of the forwarding RPL router.

P2P packets travel up toward a DODAG root, then are routed down to the final destination by the first RPL router capable of reaching the destination. In the case of "nonstoring" RPL instances packets will travel all the way to a DODAG root, which will add the source routing header (RH4 header as specified in ietf-6man-rpl-routing-header) and reinject the packet in the down direction.

12.7.1   RPL Security

RPL defines three security models:

- The "unsecured" model does not implement specific security features at RPL level, however the layer 2 network may implement some level of security (e.g., a 802.15.4 network key).
- The "preinstalled" model requires all RPL nodes to be configured with preprovisioned keys, which they use to code and decode secure RPL messages. Secure RPL messages have the high order bit of the code field set (see Figure 12.7).
- The "authenticated" mode that also uses a preinstalled key, but only to join the network as a leaf node. The node will need to obtain a key or a certificate from an authentication authority to join an authenticated RPLInstance as a router. This last mechanism is not fully defined yet and will require future companion specifications.

13

ZigBee Smart Energy 2.0

13.1 REST Overview

"Representational state transfer", or REST is a distributed software architecture style that was described by Roy Fielding in a thesis presented in 2000. The thesis discusses client-server based architectures, analyses the reasons for the success of HTTP and hypertext, and presents a number of constraints that define a RESTful architecture, that is, an architecture that will use the same design principles, and share the same desirable properties as HTTP (scalability, simplicity, reliability . . .).

13.1.1 Uniform Interfaces, REST Resources and Resource Identifiers

The first design constraint is that interfaces should be "uniform", based on the concept of exchanging *resources*. Roy Fielding introduces the concept of a resource, an abstraction for server-side information (and associated native data representation): "*Any information that can be named can be a resource*". Resources are associated to resource identifiers. A REST interface will transmit only a *representation* of such resource, which is a specific way of presenting the resource to a client that can evolve over time, or depend on the client type, while the native data representation evolves independently. Technically, a representation is "*a sequence of bytes*", plus *representation metadata* (to describe the structure and semantics of this byte sequence), and optional *resource metadata* (to represent information about the resource independent of its representation).

The REST messages exchanged between a client and a server include the identifier of the resource, an optional resource representation data, and optional control data that may indicate the purpose of a message (e.g., action being requested) or be used to parameterize the requests (e.g., select a specific representation of a resource).

It is perhaps easier to understand the specific approach of REST by comparing with other traditional programming architecture styles. Most IT systems interfaces are based

The Internet of Things: Key Applications and Protocols, First Edition.
Olivier Hersent, David Boswarthick and Omar Elloumi.
© 2012 John Wiley & Sons, Ltd. Published 2012 by John Wiley & Sons, Ltd.

on two concepts: a verb describing the action to be taken, and some form of data. This is typical of object-oriented systems, where all data is encapsulated in the server, and manipulated only through methods (verbs) and associated method parameters. Usually, the goal of such interfaces is to isolate clients from the internal data model of the server. With such verb/parameter interface models a developer needs to be familiar with two dimensions of the interface definition: the various verbs available for each interface, and the associated parameters. Reading the interface documentation is a prerequisite to coding.

In contrast, as REST interfaces are representation centric, a small set of verbs, uniform across all use cases, can be used. Usually, this set of verbs is referred to as CRUD for create, read, update and delete. Developers need to focus only on the resource representation format. This approach is very developer friendly: in many cases, knowing the resource identifier is enough to start coding. Reading the resource will provide a representation, and often representations are sufficiently self-descriptive for a developer to have a good intuition of what to do next to manipulate the resource. Many web portals provide a REST interface with very little documentation for the general public, giving access to a comprehensive feature set, and still integrating functionality from those portals "feels" easy.

An additional advantage of verb standardization is that the REST transport protocol can be specified independently of any use case, which makes it possible to define standard transport protocols (HTTP, CoAP), as well as generic application-level components (Roy calls them "*connectors*") such as proxies, load balancers, firewalls, and so on.

Of course, while this "visibility" is one of the design goals of REST, the REST constraints still provide some flexibility that can be used to defeat the original intention: for instance, it is possible to map an object-oriented interface by implementing one resource per verb, or by using one control message per verb.

At present, while REST is probably more human friendly, it is not quite as machine friendly as other interface models, such as SOAP. So far, it fails to provide a comprehensive interface description language. The Web application description language (WADL) serves that purpose, but only for a subset of potential REST interfaces. In practice, most recent RESTful architecture standards, such as oBIX, ZigBee SEP 2.0 or ETSI M2M use text specification for the definition of REST interfaces.

13.1.2   REST Verbs

The paper of Roy Fielding never stated which verbs a REST architecture had to provide, as the central idea was that those verbs should aim at manipulating resources. However, the design principles of HTTP, and its evolution to HTTP 1.1, are discussed at length, and in practice many recent standards start by specifying their HTTP binding, before considering other potential bindings for example, to CoAP or other REST-capable protocols.

The exact use of HTTP verbs in a REST context is sometimes ambiguous. Recent standards (ZigBee SEP 2.0, ETSI M2M) have converged on the following usage guidelines:

- GET: request verb for reading a representation of a resource in a *safe* fashion, that is, the request does not change the state of the resource on the server. If the resource URI represents a collection, a list of URIs of collection members will be returned. Generally, resources should be exposed according to the principle of *"gradual reveal"*, that is, structured in a way that complex structure subelements will be represented by reference in the representation of the parent resource. Another way to see this is that representations should include references to related representations, or *"hypermedia as the engine of application state"*.
- PUT: request verb for creating or replacing a resource, in an *idempotent* fashion, that is, multiple identical requests should have the same effect as a single request. If the URI represents a collection, the entire collection is replaced.
- POST: request verb for appending to a resource or creating a subordinate resource (not safe nor idempotent). For instance, a POST /item would typically result in the creation of /item/*<subresource instance number assigned by the server>* for example, /item/1. The URI of the created resource is returned in the location header as part of the 201 "created" response.
- DELETE: interface for deleting a resource (*idempotent*). If the URI represents a collection, the entire collection is replaced.
- HEAD (optional): interface to request metadata regarding a resource, in a safe fashion.
- OPTIONS: interface to request the methods available at the server for a resource, for the authorization level of the client.

13.1.3   Other REST Constraints, and What is REST After All?

The RESTful architectural style is defined by additional constraints:

- Communications should be stateless, "each request from client to server must contain all of the information necessary to understand the request, and cannot take advantage of any stored context on the server. Session state is therefore kept entirely on the client. This constraint induces the properties of visibility, reliability, and scalability. Visibility is improved because a monitoring system does not have to look beyond a single request datum in order to determine the full nature of the request. Reliability is improved because it eases the task of recovering from partial failures. Scalability is improved because not having to store state between requests allows the server component to quickly free resources [...]".
- Server responses should classify responses as cacheable or not. The use of caching improves scalability and the user experience (reduced latency).
- The interfaces should facilitate layered architectures (e.g., use intermediate load-balancing servers).

However, the core constraints are those outlined in Sections 13.1.1 and 13.1.2.

There were several recurrent discussions on the Internet and within standard bodies debating whether this or that architecture was RESTful or not. Here are some common topics:

- Are "servers" and "clients" defined at function level or interface by interface? Both ZigBee/Homeplug and ETSI concluded that many "real-world" applications cannot strictly separate functions as server or client. Many functions are both server and client for different interfaces. Therefore, the REST concepts are usually understood and applied on a per interface basis (client and server interfaces).
- What about subscribe/notify? This key functionality appears to be missing in the original REST paper, and is also missing in HTTP, leading to workarounds such as polling for AJAX interfaces. Recent standards reintroduce a subscribe/notify model by defining a dedicated resource to store subscriptions to resource R, and define a resource that need to be implemented by hosts interested in notification related to R, where notifications will be posted. This is a typical example of a case where "clients" of a resource R hosted on a server, will need to implement a server function, while the "server" will act as a client to post notifications.
- Concurrent access control to a resource. When the HTTP binding is used, the etag value is used to prevent simultaneous resource updates race conditions. All resource representations for a given resource are required to have the same etag, and the etag should be changed each time the resource is updated. Clients that want to perform a resource modification based on the assumption that the resource has not been updated since the last time they read it should include a condition based on the etag value (If-Match HTTP header).

With these clarifications, it seems that there is now a fairly good agreement among standard bodies on the practical implementation of REST style architectures. Among the work topics for further alignment:

- URI naming conventions, for example, how to represent collections;
- agreement on partial resource access/update methods, for instance using XCAP (RFC 4825) or xPath (defined by the W3C), at least for XML-encoded resources.

13.2   ZigBee SEP 2.0 Overview

The ZigBee Smart Energy Profile 2.0 protocol is the result of a joint work of the ZigBee alliance and of the HomePlug powerline Alliance, who is behind the HomePlug/AV (draft IEEE P1901) CPL standard and working on the "Green Phy" CPL standard (see Section 2.1 for more information on CPL technology).

The idea was to redesign an equivalent of ZigBee SE 1.0, but in a physical-layer-independent way, based on an IP networking layer and using a RESTful design. The clusters of ZigBee SE 1.0 are redesigned as "function sets" in SEP 2.0. In addition, the working group took into account a comprehensive requirements list coming from utility companies as well as the Society of Automotive Engineers (SAE) for aspects related to electric vehicles.

This chapter summarizes the 0.7 draft version of Smart Energy 2.0 which was published in April 2010 and is available on the web site of the ZigBee alliance: http://www.zigbee.org/Standards/ZigBeeSmartEnergy/Version20Documents.aspx.

SEP 2.0 assumes an all IPv6 network. The protocol is designed according to the REST paradigm, and the data model is intended to map directly to IEC 61968 (the "common information model"). Resources are modeled using XML, and resource representations are compressed using EXI.

As this book was going to press the draft specification was updated. The overall functionality and design remains similar, but unfortunately the data model included too many changes for us to be able to fully update this chapter. We have inserted notes in the text to outline the main changes.

13.2.1 ZigBee IP

Since SEP 2.0 relies on an IP stack, and the typical implementation targets low-powered radio device, the first step was to define an IP transport layer over for 802.15.4 networks. ZigBee decided to adopt the work of IETF 6LoWPAN and ROLL working groups and mainly focused on selecting among the various options proposed:

- It uses **IEEE 802.15.4-2006 physical and mac layers.** The non-IP version of ZigBee uses the 2003 version of 802.15.4: this update means that ZigBee IP, unlike its predecessor, will be able to run on all the frequencies supported by 802.15.4, including 900 MHz and 868 MHz, since the 2006 version introduced higher-bitrate options for these frequencies (250 kbps, up from 20 and 40 kbps, respectively, see Chapter 1).
- **6LoWPAN is used with the hc adaptation layer** (RFC 6282 "Compression Format for IPv6 Datagrams over IEEE 802.15.4-Based Networks"), and uses 6LoWPAN neighbor discovery (RFC 4861, updated by https://datatracker.ietf.org/doc/draft-ietf-6lowpan-nd/)
- The **IETF ROLL RPL routing protocol** (see Chapter 12 for more details) **is used in nonstoring mode** : source routing must be used by the 6LoWPAN DODAG root, using the new **RH4 routing header** (draft-ietf-6man-rpl-routing-header, at the time of writing).
- As mandated by RPL, IPv6 packets are injected into the RPL router network with the new **RPL option for the hop-by-hop** header (see draft-ietf-6man-rpl-option-01).

- Standard TCP and UDP transport layers are used. In practice, for constrained devices UDP is more likely to be used, in conjunction with CoAP.
- Regarding security, ZigBee/IP uses **PANA** (Protocol for Carrying Authentication for Network Access, RFC 5191 and 5193) on top of UDP to transport the EAP authentication during the client–network authentication phase for the **EAP** (RFC 3748)/ **EAP-TLS** (RFC 5216)/**TLS** (RFC 5246) security stack that use the ECC and RSA public key mechanisms and the PSK cipher suite. See Section 12.4 on security for more details.

13.2.2   ZigBee SEP 2.0 Resources

13.2.2.1   ZigBee SEP 2.0 and REST

According to the REST model, communicating entities are classified as clients or servers. SEP2.0 recognizes that in practice, most nodes are clients and servers, and that notion has a meaning only in relation to a given transaction, where the server hosts the representation of the resource being read, updated or deleted.

SEP 2.0 interacts with resources by means of the standard REST verbs for the HTTP binding: GET, PUT, POST, DELETE. A subscription mechanism has been added that can be used by nonsleepy devices (see Section 12.2.2.2).

When HTTP is used, the transport protocol is TCP on port 80 or TLS on port 443. CoAP on constrained devices will use UDP instead, however the use of CoAP was still debated by the ZigBee alliance at the time of writing. The media type used by SE 2.0 is application/exi (see Figure 13.5), which refers to the efficient XML interchange (EXI) encoding format defined by the W3C. EXI uses prior knowledge of the XML grammar of the documents exchanged between a client and a server to encode the document in a compressed format. EXI has reached "proposed recommendation" status on January 2011.

13.2.2.2   SE 2.0 Lightweight Subscription/Notification Mechanism

A device (A) that supports the subscription mechanism must expose the /sub collection resource. It will send notifications to all devices that have subscribed for event notifications by inserting a subscription subresource in the /subcollection.

A device (B) that supports receiving event notifications must expose the /ntfy resource. In order to subscribe to notifications regarding //{host A}/resource, device B must send a POST to //{host A}/sub/{IPv6 address of B} containing the URI of the monitored resource (//{host A}/resource) and the URI of its own notification resource (//{host B}/ntfy), as in the example of Figure 13.1.

It is possible to list all active subscribers of (A) by reading /sub. Figure 13.2 shows an example of the /sub resource representation (as for other examples of this chapter, the resource representation is shown decoded, it would actually be EXI encoded).

```
PUT /sub/{IPv6 Address of A} HTTP/1.1
Host: {IPv6 Address of B}
Content-Type: application/exi

<?xml version='1.0' encoding='UTF-8'?>
<SubscriptionList xmlns='http{s}://www.zigbee.org/doc/se-2-0-0'>
    <Subscription>
        <Resource>http{s}://{IPv6 Address of A}/resource
            </Resource>
        <NotificationURI>http{s}://{IPv6 Address of B}/ntfy
            </NotificationURI>
    </Subscription>
    <Subscription>
        <Resource>http{s}://{IPv6 Address of A}/resource2
            </Resource>
        <NotificationURI>http{s}://{IPv6 Address of B}/ntfy
            </NotificationURI>
    </Subscription>
</SubscriptionList>
```

Figure 13.1   SE 2.0 notification subscription example.

If (A) detects a change in the monitored resource, it will send a POST to //{host B}/ntfy containing the URI of the modified resource, and the URI of (B)'s subscription (to be used as a handle to the trigger subscription, or simply a reminder of the URI that can be used to change the subscription). Figure 13.3 shows an example notification.

Note: As this book was going to press, an update of the draft specification introduced a EndDevice list resource where a server stores data related to each client. A subresource (e.g. edev/1) is created for each client. The client subscriptions are now stored in edev/{#}/sub.

```
<?xml version='1.0' encoding='UTF-8'?>
<SubscriberList xmlns='http{s}://www.zigbee.org/doc/se-2-0-0'>
    <Subscriber href="http{s}://{IPv6 Address}/sub/{IPv6
        Address 1}" name="{IPv6 Address 1}" />
    <Subscriber href="http{s}://{IPv6 Address}/sub/{IPv6
        Address 2}" name="{IPv6 Address 2}" />
</SubscriberList>
```

Figure 13.2   SE 2.0, example list of subscribers (/sub).

```
POST /ntfy HTTP/1.1
Host: {IPv6 Address of A}
Content-Type: application/exi

<?xml version='1.0' encoding='UTF-8'?>
<Notification xmlns='http{s}://www.zibee.org/doc/se-2-0-0'>
    <Resource>http{s}://{IPv6 Address of A}/resource4</Resource>
    <SubscriptionURI<http{s}://{IPv6 Address of B}/SUBSCR1BE/
{IPv6 Address of A}
    </SubscriptionURI>
</Notification>
```

Figure 13.3   Example SE 2.0 notification posted to /ntfy.

13.2.2.3   SE 2.0 Collection and Event Resources

Figure 13.4 lists the conventions used by SE 2.0 to represent collection resources and resource instances that are part of a collection. Events are stored as a collection, with a specific alias to the active event.

URI http(s): //{address}/...	Description
/{path to a collection of events}/act	Alias of the currently active event resource
/mrr	List of links to mirrored feature sets, e.g. /mrr/0

/egg	Collection of 'egg' resources, GET /egg typically returns :
	
	 `<EggList xmlns='http{s}://www.zigbee.org/doc/`
	 `se-2-0-0'>`
	 `<Egg href="http{s}://{IPv6 Address}/egg/0"`
	 `name="0" />`
	 `<Egg href="http{s}://{IPv6 Address}/egg/1"`
	 `name="1" />`
	 `</EggList>`
	 POST /egg creates a new egg instance e.g. /egg/1. PUT is not allowed on collections. DELETE /egg deletes all sub-resources of /egg.

/egg/1	Egg resource instance # 1, as part of a collection. PUT /egg/1 replaces this instance or creates it if it did not exist.

Figure 13.4   SE 2.0 collection and event resources.

13.2.2.4 Resource Discovery, /rsc

For the resolution of host names, device and resource discovery, SE 2.0 uses mDNS (IETF draft-cheshire-dnsext-multicastdns) and DNS-SD (draft-cheshire-dnsext-dns-sd). DNS service discovery uses DNS PTR records mapping a <Service>.<Domain> (e.g. Test\032Server._smartenergy._tcp.local.) name to a list of <Instance>.<Service>.<Domain> names. Each <Instance>.<Service>.<Domain> can be located by using DNS SRV records, and additional service information is stored in TXT records in key/value pair format. The URI of an Smart Energy 2.0 DeviceCapabilities resource is stored in the path key (e.g. path=/dcap).

The list of resources on a given host (identified by its IP address) is represented at URI /rsc, and enumerates all logical device types supported, and the URI of the related function set (see Section 12.3). In ZigBee 1.0 terms, this is equivalent to the list of "endpoints", each endpoint corresponding to a function set.

The list may also contain mirrored resources for other devices (e.g., sleeping or mobile devices). The list of mirrored resources may also be found at /mrr.

13.3 Function Sets and Device Types

SE 2.0 defines a "function set" as a group of related functionalities (the equivalent of a cluster in ZigBee 1.0). Each function set defines a list of REST resources and associated transactions, and is identified by a resource name. The resource name is used for the resource discovery mechanism.

SE 2.0 defines the following function sets:

* demand response/load control;
* messaging;
* confirmation;
* pricing;
* prepayment;
* metering;
* plug-in electric vehicles;
* distributed energy resource;
* billing;
* registration;
* base;
* device management/configuration;
* firmware download server;
* firmware download client;
* diagnostics and monitoring.

```
HTTP/1.1 200 OK
Content-Type: application/exi

<?xml version='1.0' encoding='UTF-8'?>
<DeviceList xmlns='http{s}://www.zigbee.org/doc/se-2-0-0'>
    <Common>
        <Profile>Common</Profile>
        <Basic href="http{s}://{IPv6 Address}/Basic/0" name="0"/>
    </Common>
    <MeteringDevice>
        <Profile>Smart Energy</Profile>
        <Type>Electric</Type>
        <Meter href="http{s}://{IPv6 Address}/Meter/0" name="0"/>
    </MeteringDevice>

    <ESI>
        <Profile>Smart Energy</Profile>
        <Time href="http{s}://{IPv6 Address}/Tnme/0" name="0"/>
    </ESI>

    <ThermostatDevice>
        <Profile>Home Automation</Profile>
        <Thermostat href="http{s}://{IPv6 Address}/TSTAT/0"
            name="0"/>
    </ThermostatDevice>
</DeviceList>
```

Figure 13.5 Example response to GET /rsc (data represented in bold is actually compressed using EXI).

Figure 13.6 lists the device types that are defined in SE 2.0. Each device type is characterized by the function sets it must implement on the client or the server side. This is a transposition of the model of ZigBee 1.0, where each device type was characterized by the supported clusters on the server and client sides.

13.3.1   Base Function Set

The base function set groups general support resources that are useful to most end devices. Some of them can also be implemented by the ESI (e.g., the time resource). The resources currently defined as part of the base function set are briefly described in Figure 13.7 and Figure 13.8. The usage of most of them is self-explanatory.

The usage for the randomize resource and the power configuration resources is described in more detail in the following sections.

Device type	Mandatory function set client or server
In Premises Display	Client of the Metering, Price, or Message function sets.
Load Control	Client of the Demand Response Load Control function set.
Smart Thermostats	Client of the Demand Response Load Control or Price function set.
Meters	Server of the Metering function set
Smart Appliances	Client of the Demand Response Load Control or Price function set.
Premises Energy Management Systems	Both a client and a server of either the Demand Response Load Control or Price function sets
Energy Services Interface	Server of the Message, Price, and Demand Response Load Control Function Sets.
Prepayment Terminals	Client of the Price and Billing function sets
Inverters	Server of the DER function set.
Electric Vehicle Supply Equipment (EVSE)	Client of the Plug-In Electric Vehicle List, Price, and Demand Response Load Control function sets, and a server of the Plug-In Electric Vehicle.
End Use Measurement Device (EUMD)	Server of the Metering function set

Figure 13.6   ZigBee SEP 2.0 device types and mandatory function sets.

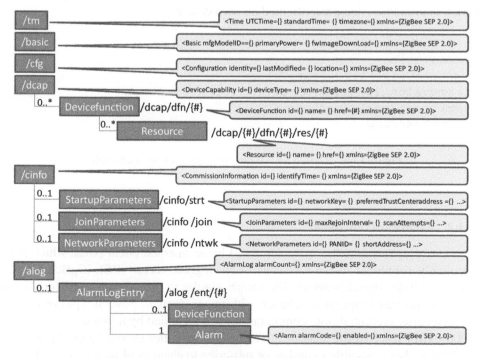

Figure 13.7   ZigBee SEP 2.0 base resources (part 1).

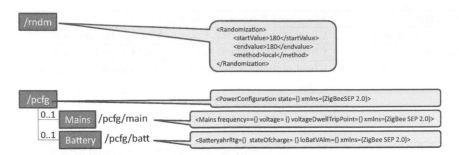

Figure 13.8   ZigBee SEP 2.0 base resources (part 2).

13.3.1.1   Randomization

Telecom service providers know the potential catastrophic effects of massive synchronization of client devices (see Appendix B). In order to provide a generic tool to avoid these situations, ZigBee SE 2.0 provides a randomization mechanism that can be used by any function set potentially affected by this problem: at present, the demand response load control, and price function sets.

Randomization is supported through the /rndm resource (see Figure 13.8 and Figure 13.9). Depending on the service-provider policy, the implementation may rely on the effective calculation of a random value, within the specified bounds, by the device, or on a fixed random value preconfigured on each device.

```
<Randomization>
    <startValue>180</startValue>
    <endvalue>180</endvalue>
    <method>local</method>
</Randomization>
```

Figure 13.9   ZigBee SE 2.0 randomization resource example.

13.3.1.2   Firmware Download

The firmware upload function is implemented by an upgrade client (UC), which polls, or optionally subscribes to the resources of an upgrade server (US). ZigBee SEP 2.0 defines a digitally signed firmware file format similar to that already used for ZigBee 1.0 over-the-air (OTA) upgrading cluster, with minor changes. Several types of files are defined (security credential, log, configuration) and identified by a two-octet value, with $0\times0000\text{-}0x\text{ffbf}$ reserved for manufacturer use (one identifier should be used per device type). This makes it possible to update or subscribe to changes of only certain file types ($0x\text{ffff}$ is defined as the wildcard file type).

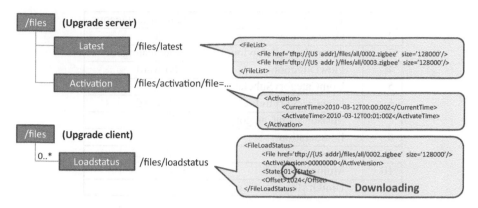

Figure 13.10 ZigBee SEP 2.0 firmware upload resources.

On the upgrade server, the URI/files/latest returns the list of new files targeted for the client. The request optionally specifies the manufacturer code, file type and current file version. Deferred activation is supported by means of the Activation resource. The upgrade server resources are outlined in Figure 13.10.

On the client side, the status of configuration files can be verified by reading the /files/loadstatus resource.

13.3.2   Group Enrollment

Energy service providers usually want to direct their commands (e.g., demand-response commands) to only a subset of all potential controlled devices. This might be to avoid synchronization effects, to control the aggregate energy volume affected by the command, or to scope the command geographically. SE 2.0 supports these requirements by providing the notion of a group enrollment resource.

The SE 2.0 group enrollment URI is /enrl.

```
<?xml version='1.0' encoding='UTF-8'?>
<EnrollmentList xmlns='http{s}://www.zigbee.org/doc/se-2-0-0'>
    <EnrollGroup>
        <Group>0</Group>
        <Group>1</Group>
    </EnrollGroup>
</EnrollmentList>
```

As this book was going to press, an update of the draft specification was published. Device group enrollments are now configured as a subresource of the EndDevice resource instance on the ESI. Each group contains links to the specific function set instances applicable to the EndDeviceGroup. /enrl resource is no longer used.

1	TimeAttribute	12 = instantaneous
2	DataQualifier	0=N/A
3	AccumlationBehaviour	6= indicating
4	FlowDirection	l=Forward
5	UomCategorySubclass	0=N/A
6	UomCategoryIndex	8=demand
7	MeasurementCategory	0=N/A
8	Enumeration	
9	Phase	0=n/a to all phases
10	Multiplier	3=kilo
11	UnitOfMeasure	38=W

Figure 13.11   IEC TC57 61968 ReadingTypeID structure.

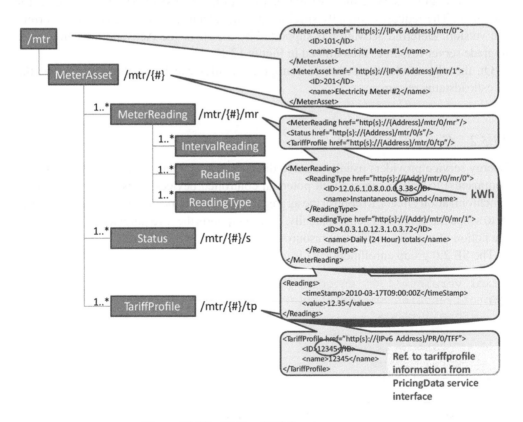

Figure 13.12   ZigBee SE 2.0 meter resources.

13.3.3   Meter

The meter structures of ZigBee SEP 2.0 have been designed in close coordination with IEC TC57 61968 and captures a baseline metering functionality. SEP 2.0 does not implement, however, more advanced functions such as programmable autoreads. The structure of the metering resources is outlined in Figure 13.12.

The ReadingTypeID format is imported from IEC TC57 61968, it is a concatenation of 9 attributes each represented by one or two integers for a total of 11 integers (Figure 13.11). For instance the present maximum indicating forward water (m³/h) is represented by 15.8.6.1.0.63.0.0.0.0.121.

13.3.4   Pricing

The price resources allow utilities to publish a description of their tariff structures on an ESI. Figure 13.13 illustrates the relationships between price resources and shows the "well-known" URIs defined by SE 2.0.

The ZigBee SE 2.0 pricing resources support both time of use (ToU) pricing and consumption interval pricing, in any combination. The example of Figure 13.14, adapted from draft 0.7, shows how time-based pricing and volume-based pricing can coexist.

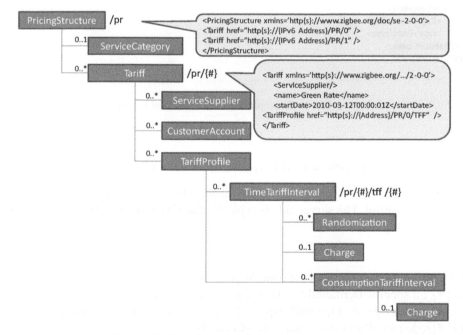

Figure 13.13   ZigBee SE 2.0 pricing resources.

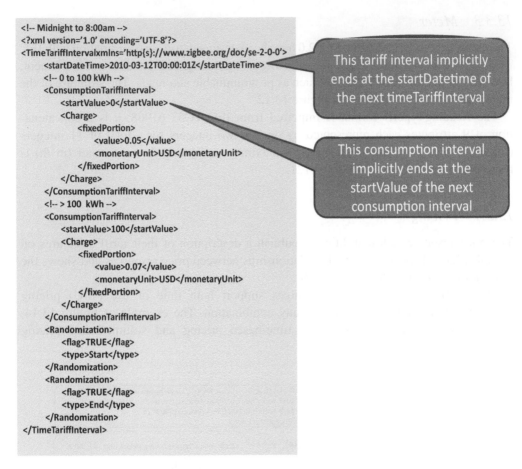

Figure 13.14   Example TimeTariffInterval resource.

13.3.5   Demand Response and Load Control Function Set

The clients of the DR/LC function set are typically smart thermostats or any device that supports load control. The server side is typically the ESI and implements the following resources (see Figure 13.15):

- /dr : a collection of DemandResponseProgram collections;
- /dr/{#} : a specific DemandResponseProgram collection resource;
- /dr/{#}/nm : a specific DemandResponseProgram name attribute resource;
- /dr/{#}/edc : a collection of EndDevicesControls;
- /dr/{#}/edc /{#} : a specific EndDevicesControl resource;
- /dr/{#}/edc /act : the active EndDevicesControl resource.

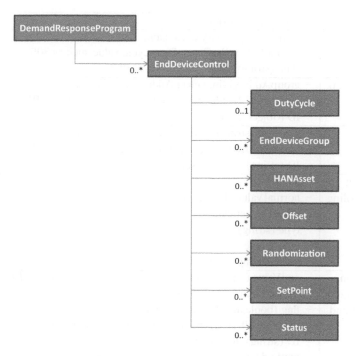

Figure 13.15 ZigBee SE 2.0 DR/LC server resources.

The EndDeviceControl resource contains or points to all the information required by a client to implement a particular dr/lc event for a period of time.

In addition to the attributes listed in Figure 13.16, each EndDeviceControl resource may also contain the subresources listed in Figure 13.17.

DR/LC attributes	Description
ProgramLevel (integer)	Level of a demand response program request, where 0=emergency.
drProgramMandatory (boolean)	Whether a demand response program request is mandatory
duration (Minutes)	Event duration (end - start)
href (anyURI)	Hypertext reference pointing to a URI
ID (string)	Object identifier
name (string)	Name of the EndDeviceControl resource
scheduledInterval (DateTimeInterval)	(if control has scheduled duration) Date and time interval the control has been scheduled to execute within.
type (string)	Type

Figure 13.16 EndDeviceControl attributes.

Sub-Resource	Sub-Resource attributes
DutyCycle	– name (string): Duty cycle name – normalValue (PerCent): Duty cycle value such as 80% – state (string): State such as on or off
EndDeviceGroup	– groupAddress (integer) : Address of this end device group. – href (anyURI) : Hypertext reference pointing to a URI – ID (object identifier)
HANAsset	– category (string): Utility-specific categorization of this document. – href (anyURI) : Hypertext reference pointing to a URI – ID (string) : HAN asset identifier
Offset : Offset such as cooling or heating offset	– name (string) – normalValue (PerCent) – Offset as per cent – type (string) – Offset type – value (string) : offset value
Randomization : Randomization for start or end of an event	– endValue (unsignedInt) : End randomization value in SEP UTCTime format such 300 for 5 minutes – flag (boolean) : Randomization or not – href (anyURI) : Hypertext reference pointing to a URI – ID (string): Identifier – method (string) – Local (maximum) or static randomization – name (string) : Randomization name – startValue (unsignedInt) : Start randomization value in SEP UTCTime format such 300 for 5 min – type (string) : Randomization type (start or end randomization)
SetPoint : A SetPoint is an analog control used for supervisory control.	– maxValue (float) : Normal value range maximum for any of the Control.value. Used for scaling, e.g. in bar graphs. – minValue (float) : Normal value range minimum for any of the Control.value. Used for scaling, e.g. in bar graphs. – name attribute (string) : Name of an attribute. – normalValue (float) : Normal value for Control.value e.g. used for percentage scaling – value (float) : Value in type of float
Status: Current status information relevant to an entity.	– dateTime (unsignedInt) : Date and time for which status 'value' applies. – href (anyURI) : Hypertext reference pointing to a URI – reason: Reason code or explanation for why an object went to the current status 'Value'. – value (string) : Value in string

Figure 13.17   DR/LC EndDeviceControl subresources.

The DR client would typically send the following request to the ESI:

```
GET /esi HTTP/1.1
Host: {IPv6 Address}
```

The server would respond:

```
HTTP/1.1 200 OK
Content-Type: application/exi
<?xml version=''1.0'' encoding=''UTF-8''?>
<EndDeviceControl href='' http{s}://{IPv6 Address}/dr/0/edc/0''>
<ID>101</ID>
<Randomization>
<flag>TRUE</flag>
<type>Start</type>
</Randomization>
<Randomization>
<flag>TRUE</flag>
<type>End</type>
</Randomization>
</EndDeviceControl>
```

13.3.6 Distributed Energy Resources

The ZigBee S2.0 SEP resources that model distributed energy resources are outlined in Figure 13.18.

The power-generation curve is perhaps the most important piece of information from a utility perspective. This resource stores the stepwise linear approximation of the power-generation curve, an example is given in Figure 13.19.

The DERstatus object contains a description of the operational state of the generating unit, such as "starting up", the number of times the generator has been started since the last counter reset, and the total operation time of the generator since the last counter reset.

The DERcontrol object allows a utility to offset the production level to a certain percentage, and includes options to randomize such commands.

13.3.7 Plug-In Electric Vehicle

One of the priority action plans (PAP 11[1]) of the US National Institute of Standards and Technology (NIST) Smart Grid Interoperability Panel (SGIP) is to design common object models for electric transportation. PAP11 formulates the issue as follows:

---

[1] http://collaborate.nist.gov/twiki-sggrid/bin/view/SmartGrid/PAP11PEV.

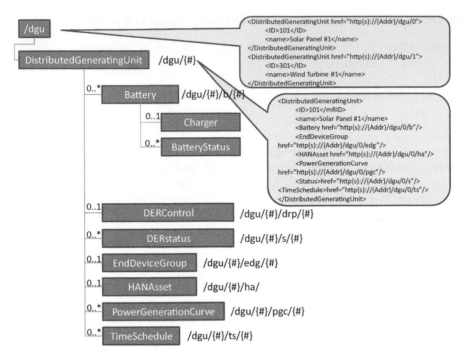

Figure 13.18   ZigBee 2.0 SEP distributed energy resources data model.

```
<PowerGenerationCurve href="http{s}://{IPv6 Address}/dgu/0/pgc/0">
    <curveType></curveType>
    <description></description>
    <ID></ID>
    <xUnit>s</xUnit>
    <y1Unit>W</y1Unit>
    <CurveData>
        <xvalue>1.0</xvalue>
        <y1value>12.34</y1value>
    </CurveData>
    <CurveData>
        <xvalue>2.0</xvalue>
        <y1value>42.34</y1value>
    </CurveData>
    <CurveData>
        <xvalue>3.0</xvalue>
        <y1value>72.34</y1value>
    </CurveData>
</PowerGenerationCurve>
```

Figure 13.19   ZigBee 2.0 SEP PowerGenerationCurve example.

The introduction of mobile plug-in electric vehicles (PEVs) to the grid creates some interoperability challenges around exchanging price, demand response (DR), and settlement information. The impact of PEVs on the grid is expected to be significant, and the ability to control the charging profiles through price or direct control, the need for cyber security (including appropriate privacy), the issues of safety, the possibility of allowing customers to sell PEV electricity back into the grid, and complexity of providing fair settlement to everyone in the value chain when vehicles charge away from their home base, requires common object models to manage all these aspects.

As of March 2011, no firm decision had been made by the SGIP governing board to adopt the ZigBee SE 2.0 information model (repackaged as SAE J2847/1) as a standard, however, it was the most likely candidate to be adopted during the ongoing standardization process to be held within the SGIP Vehicle to Grid (V2G) working group,[2] and within IEC TC 57 (ZigBee SE 2.0 PEV and DER information model are designed as an extension of the common information model, or CIM, defined in IEC 61968 and IEC 61850). For more background information on EV charging, refer to Chapter 6.

The structure of the ZigBee 2.0 SEP resources that compose the PEV function set is outlined in Figure 13.20.

The ElectricVehicle object is characterized by the objects listed in Figure 13.20, and following attributes:

- **ID**: a string;
- **odometerReadDateTime**: an unsigned integer indicating the date of the odometer[3] reading;
- **odometerReading**: a string indicating the actual reading;
- **status**: an indicator of the status of the battery, which includes indicators for over and under charge conditions, and a per cent indicator of the state of charge.

The battery object indicates the battery technology (batTyp attribute), Ah capacity rating of the battery (ahrRtg), and battery nominal voltage.

A specific charger resource is designed to retrieve the battery charge information: http{s}://{IPv6 Address}/pev/{#}/bttr/{#}/ct where "ct" stands for "charge transaction". The charger resource indicates, among other things, the active power charge rate (reChaRte attribute) of the battery, and the charging schedule (start and end time).

The battery status subresource: http{s}://{IPv6 Address}/pev/{#}/bttr/{#}/s includes indicators for over- and undercharge conditions, and a per cent indicator of the state of charge for the specific battery.

The DemandResponseProgram collection contains a list of pointers to DemandResponsePrograms, identified by their ID (the actual DemandResponseProgram resource resides on the ESI), and applying to this PEV.

---

[2] http://collaborate.nist.gov/twiki-sggrid/bin/view/SmartGrid/V2G.
[3] "An odometer indicates distance traveled by a car or other vehicle" (Wikipedia).

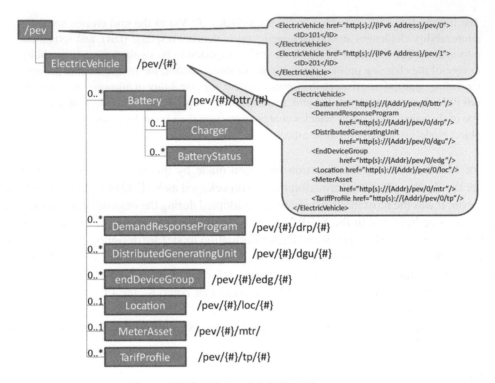

Figure 13.20   ZigBee 2.0 SEP PEV resources.

13.3.8   Messaging

The messaging function set uses two types of resources:

- TextMessages (see Figure 13.21) is typically implemented in the ESI and must be polled or subscribed to by the clients. The resource URI http{s}://{server IPv6 Address}/msg/ provides access to a collection of message collections. Individual messages are accessed by their individual URIs http{s}://{server IPv6 Address}/msg/{#}/txt/{#}. The structure of each TextMessage, as illustrated on Figure 13.21, enables the ESI to display each message to only certain groups of devices, or to certain device types, at during certain periods of time. Each message may request a confirmation.
- Confirmations. That resource is implemented (or mirrored to a parent device) by in premises displays that are capable of message confirmation. The resource URI http{s}://{IPv6 Address}/rsp/ provides access to a collection of responses. The structure of the confirmation resource includes the confirmed message ID and a time stamp. It can also be used to confirm prices or billing.

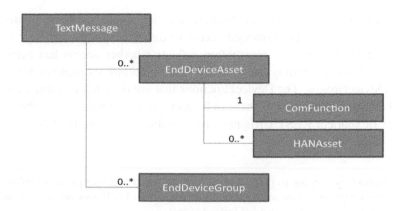

Figure 13.21 ZigBee SE 2.0 TextMessage resource.

13.3.9 Registration

The Registration function set and associated resources (Figure 13.22) are used to define which devices are authorized to access which function sets.

The information related to each end device is stored in the HANAsset resource and subresources. Keeping the harmonization with the IEC Common Information Model, the HANAsset class is a type of CIM "EndDeviceAsset".

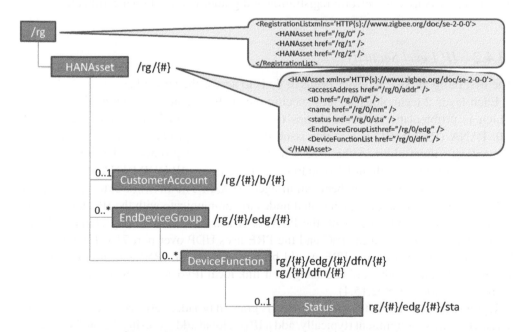

Figure 13.22 ZigBee SEP 2.0 registration resources.

The DeviceFunctions list the FunctionSets that the device has been granted access to prior to registration, or has requested access to through the registration process. The "status" of the DeviceFunction registration reflects whether access has been granted or if the request is still pending. An end device may declare its membership in one of several EndDeviceGroups. The DeviceFunctions that are declared as subresources of an EndDeviceGroup are function sets that can be accessed by all members of the group.

A number of shortcut REST URIs have been defined to access HANAsset attributes, which are listed in Figure 13.23.

Note: As this book was going to press, an update of the draft specification was published. The registration of each device is now materialized by an instance of EndDevice resource on the server to which the device registered, e.g. /edev/1 instead of rg/1.

13.4   ZigBee SE 2.0 Security

13.4.1   Certificates

A device certificate is installed during manufacturing on each SE device, and is not expected to change during the lifetime of the device (as long as the network ID, e.g., EUI-64, of the device does not change).

An operational certificate is used to secure a given session or relationship. The registration server is used to perform registration and grants an operational certificate.

13.4.2   IP Level Security

The first thing an SE 2.0 device needs to do is to access the network.

Each layer 2 technology defines specific means for the authentication of new nodes, which is problematic for technologies that want to remain layer 2 agnostic, like SE 2.0. PANA (RFC 5191) solves this issue by defining a protocol that allows clients to authenticate themselves to the access network using IP protocols. The PANA client (PaC) is configured with an IPv6 address, and exchanges IP UDP PANA authentication messages with the PANA authentication agent (PAA). The network enforcement point (EP) is supposed to let unauthenticated nodes to communicate with the IP address of the PAA : it may be collocated with the PAA, or it acts as a PANA relay element (PRE). Communication between the PaC and the PRE uses UDP over port 716. Typically, on a 802.15.4 radio network, the PRE would be the parent node (PN) and the joining node would be the PaC, both would be using their link local IPv6 addresses (see Chapter 1 for more information on 802.15.4).

Upon successful authentication, the PaC is granted broader network access possibly by a new IP address assignment (typically, add a IPv6 global address using the prefix obtained during earlier IPv6 router discovery), by enforcement points changing filtering rules for

/rg/{#}/addr	Access address for a specific HANAsset object. E.g. : `<HANAsset accessAddress={Addr} xmlns=' HTTP{s}://www.zigbee.org/doc/se-2-0-0' />`
/rg/{#}/id	unique resource id for a specific HANAsset object.
/rg/{#}/nm	name of a specific HANAsset object E.g: `<HANAsset name="Device0" xmlns='HTTP{s}: //www.zigbee.org/doc/se-2-0-0' />`
/rg/{#}/sta	status value for a specific HANAsset object.
/rg/{#}/edg	is a collection of EndDeviceGroup objects. E.g.: `<EndDeviceGroupList xmlns='HTTP{s}://www. zigbee.org/doc/se-2-0-0'>` `<EndDeviceGroup href="/rg/0/edg/0" />` `<EndDeviceGroup href="/rg/0/edg/1" />` `</EndDeviceGroupList>`
/rg/{#}/edg/{#}	specific EndDeviceGroup object.
/rg/{#}/edg/{#}/addr	Group address of a specific EndDeviceGroup object. `<EndDeviceGroup groupAddress={IPv6 Address} xmlns=' HTTP{s}://{ZigBee SEP}'/>`
/rg/{#}/edg/{#}/id	unique resource id for a specific EndDeviceGroup object.
/rg/{#}/edg/{#}/dfn	Collection of DeviceFunction objects associated with a specific EndDeviceGroup object. `<DeviceFunctionList xmlns='HTTP{s}://www. zigbee.org/doc/se-2-0-0'>` `<DeviceFunction href="/rg/0/edg/0/ dfn/0" />` `<DeviceFunction href="/rg/0/edg/0/ dfn/1" />` `</DeviceFunctionList>`
/rg/{#}/edg/{#}/dfn/{#}	specific DeviceFunction object associated with an EndDeviceGroup
/rg/{#}/edg/{#}/dfn/{#}/id	unique resource id for a specific DeviceFunction object associated with an EndDeviceGroup.
/rg/{#}/edg/{#}/dfn/{#}/id	unique resource id for a specific DeviceFunction object associated with an EndDeviceGroup.
/rg/{#}/edg/{#}/dfn/{#}/nm	name of a specific DeviceFunction object associated with an EndDeviceGroup.
/rg/{#}/edg/{#}/dfn/{#}/sta	Registration status for a specific HANAsset, DeviceFunction, and EndDeviceGroup.

Figure 13.23 ZigBee SEP 2.0 registration resource URIs.

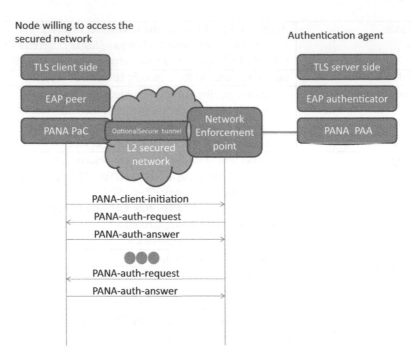

Figure 13.24   Overview of PANA.

the same IP address, or in the specific case of 802.15.4 by obtaining the network key. Extensions to PANA are currently being introduced to support such group key distribution (draft-ohba-pana-keywrap-02 as of March 2011).

Other than that, PANA simply encapsulates EAP payload, which enables use of any authentication method supported by EAP. Figure 13.24 illustrates the security stack used to access the network, and the use of PANA.

Integrity protection of messages between the PaC and PAA is possible as soon as EAP exchanges have generated a secure association (SA) shared key. In addition PANA can establish an optional IPsec tunnel to the PAA if the layer two is not secure and encryption needs to be provided.

ZigBee SE 2.0 uses EAP-TLS (RFC 5216), an EAP transport method for transport layer security (TLS). It uses TLS 1.2 (RFC 5246), and mandates support for the cipher suites TLS_PSK_WITH_AES_128_CCM_8 and TLS_ECDHE_ECDSA_WITH_AES_128_CCM_8. The AES-CCM cipher is implemented in many 802.15.4 chipsets in hardware. The standard TLS handshake is used to provide mutual authentication and to derive a shared security association.

The ZigBee *authentication server* manages incoming EAP-TLS transactions from joining nodes, performs network authentication and controls network access. It is typically hosted by the ESI. Its functional role is roughly equivalent to that of the application trust center in ZigBee 1.0.

13.4.3 Application-Level Security

13.4.3.1 Registration

Once the device is able to communicate to other nodes on the IP network, it still needs to register to the utility or service provider registration server (e.g., the ESI) in order to be able to access the service provider SE 2.0 resources.

The device authenticates with the registration server using TLS handshake based on device certificates, and the registration server grants one or more operational certificates to the device (using the TLS record protocol): one operational certificate is associated to the device, and operational certificates are associated to individual resources on the device. At the application level, ZigBee SE 2.0 uses TLS cipher suites TLS_DHE_RSA_WITH _AES_128_GCM_SHA256 and TLS_ECDHE_ECDSA_WITH _1854 AES_128_ GCM_SHA256, and X.509v3 certificates. A device may register with multiple registration servers (e.g., multiple utilities or service providers).

13.4.3.2 Authorization Server, ACLs

Any device that contains protected resources must also implement the authorization server function: it authenticates the client requesting access to the resource using TLS and the operational certificates, and then uses TLS negotiation to establish a secure tunnel. Therefore, in addition to native local network security, ZigBee SE 2.0 provides application-level secure tunnels. These secure tunnels are specific to each set of operational certificates, therefore to each utility registration: multiple utilities may securely share the same SE 2.0 network. Once the identity of the client has been asserted, the authorization server then uses application-defined ACLs to restrict access to the protected resources according to its security policy.

Privilege	Description
GET	allowed to perform the GET method on the resource
PUT	allowed to perform the PUT method on the resource
POST	allowed to perform the POST method on the resource
DELETE	allowed to perform the DELETE method on the resource
GET_ACL	allowed to perform the GET method on the acl subresource
PUT_ACL	allowed to perform the PUT method on the acl subresource
POST_ACL	allowed to perform the POST method on the acl subresource
DELETE_ACL	allowed to perform the DELETE method on the acl subresource

Figure 13.25   ACL privileges defined in ZigBee SE 2.0.

```
<AccessControlList xmlns='http{s}://www.zigbee.org/doc/se-2-0-0'>
    <Grant>
        <ID>{IPv6 Address}</ID>
        <Privilege>GET_ACL</Privilege>
    </Grant>
    <Grant>
        <ID>{IPv6 Address2}</ID>
        <Privilege>GET</Privileg>
        <Privilege>PUT</Privilege>
        <Privilege>POST</Privilege>
        <Privilege>DELETE</Privilege>
        <Privilege>GET_ACL</Privilege>
        <Privilege>PUT_ACL</Privilege>
        <Privilege>POST_ACL</Privilege>
        <Privilege>DELETE_ACL</Privilege>
    </Grant>
    <!-- Grant GET Access to ULAs -->
    <Grant>
        <ID>fc00: :/7</ID>
        <Privilege>GET</Privilege>
    </Grant>
</AccessControlList>
```

Figure 13.26   Example ZigBee SE 2.0 ACL.

SE 2.0 ACLs are designed as white lists: access to a resource via a REST method is not granted unless an ACL explicitly allows it. The ACL of a resource is defined as subresource //{host address}/{resource}/acl, and lists which privileges are granted to each client host (see Figure 13.25).

An example ACL resource is provided in Figure 13.26.

14

The ETSI M2M Architecture

14.1 Introduction to ETSI TC M2M

At present, there are about 50 to 70 billion "machines" in the world, about 1% of which are connected to a communication network. There is obviously an enormous growth potential for M2M, but the transition from current midscale M2M applications (about 500 000 devices) to the next level (applications managing tens of millions of devices) will require new standards.

While current M2M standards address the transport level, and client to server communication protocols, the future "Internet of Things" will require a system-level architecture:

- Enabling application developers to focus on functionality, not lower-level tasks like network access control, authentication or routing;
- Enabling any application to read or control any sensor, under control of a horizontal security framework;
- Providing network-based services, such as data publication and subscription.

In order to achieve these goals, common functions and network elements need to be identified and standardized at part of the M2M infrastructure: the ETSI M2M technical committee was created in January 2009 at the request of many telecom operators to create a standard system-level architecture for mass-scale M2M. ETSI TC M2M does not address one domain in particular; on the contrary, its ambition is to become the common backbone of all mass-scale M2M applications. The following domains are explicitly covered:

- Security/serenity: surveillance applications, alarms, object/people tracking;
- Transportation: fleet management, road safety;
- Health care: personal security, e-health;
- Smart energy: measurement, provisioning and billing of utilities;
- Supply and provisioning: freight supply and distribution monitoring, vending machines;
- City automation: public lighting management, waste management;
- Manufacturing: production chain monitoring and automation.

The Internet of Things: Key Applications and Protocols, First Edition.
Olivier Hersent, David Boswarthick and Omar Elloumi.
© 2012 John Wiley & Sons, Ltd. Published 2012 by John Wiley & Sons, Ltd.

As for all recent automation protocols, the ETSI M2M architecture is resource centric and adopts the RESTful style (refer to Section 13.1 for more details on REST). As usual, the four basic verbs of REST (create, read, update, delete) are complemented at the functional level by execute, subscribe and notify primitives, which are implemented, at a lower level, by helper resources manipulated by the CRUD verbs. See Section 14.3.6 for example.

We expect ETSI M2M to become the system-level architecture of the Internet of Things, much in the same way as GSM or UMTS (also from ETSI) have become the dominant system-level architectures for mobile communications.

ETSI M2M does not aim at replacing existing standard or proprietary automation protocols, such as those described in the other chapters of this book. It aims at integrating all of these protocols into a common architecture, facilitating access to any of these vertical protocols and networks from any hosted service, in an operator-controlled way. The companion book "M2M Communications: A Systems Approach" (David Boswarthick, Omar Elloumi, Olivier Hersent) describes ETSI M2M general architecture in more details. In this chapter we will just give an overview of ETSI M2M architecture, and explain in more detail how ETSI M2M can integrate existing automation and metering protocols, using the example of ZigBee 1.0, DLMS and C.12.

14.2   System Architecture

14.2.1   High-Level Architecture

The ETSI M2M functional architecture is presented in ETSI TS 102 690. The ETSI M2M system architecture separates the M2M device domain and the network and applications domain (Figure 14.1):

- The *device domain* is composed of *M2M devices* and *M2M gateways*. ETSI M2M devices can connect to the M2M network domain directly (D' devices) or via M2M gateways acting as a network proxy (D devices). M2M gateways can be cascaded, or operate in parallel mode (e.g. for redundancy purposes).
- The *network and applications domain* comprises:
   - The access and transport network (e.g., an xDSL access network and an IP transport network).
   - The *M2M core*, which itself is composed of:
     - a *Core network* (which provides IP connectivity, service and network control functions, network to network interconnect and roaming support); and
     - *M2M service capabilities*, the functional modules implementing the M2M functions shared by multiple applications through open interfaces.
   - The *M2M applications* that run the M2M service logic and use the M2M service capabilities. The ETSI M2M architecture supports multiagent applications, which can have components running in the end devices (device application or DA), in the gateways (gateway application or GA) and in the network (network application or NA).

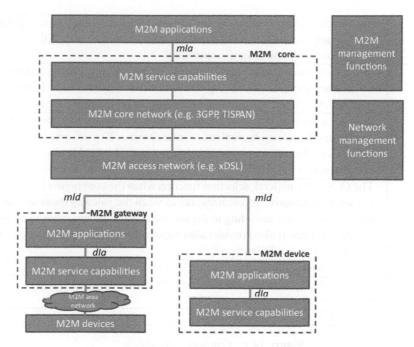

Figure 14.1   ETSI M2M high-level architecture.

- The network management functions address the access, transport and core network provisioning and supervision.
- The M2M management functions provide the services required for M2M service bootstrapping (M2M service bootstrap function or MSBF), and M2M security (M2M authentication server or MAS).

14.2.2   Reference Points

TS 102-690 defines three reference points:

- **mIa** between a M2M Network Application (NA) and the M2M service capabilities in the networks and applications domain. It provides registration and authorization primitives for the NA, service session management (event reporting or streaming sessions), and read/write/execute/subscribe/notify primitives for objects or groups of objects residing in M2M devices or gateways, as well as group objects managed by the network-domain capabilities.
- **dIa** between:

  (a) a device application (DA) and M2M service capabilities in the same M2M device or in a M2M gateway;
  (b) a gateway application (GA) and M2M service capabilities in the same M2M gateway.

GAE	This function registers gateway applications and is the single contact point for gateway applications via interface dIa, it hides the service capabilities topology and performs routing towards capabilities.
GGC	Manages the secure transport session establishment and policing according to the messages service class, using GSEC to retrieve session keys. Performs routing between Service Capabilities and the NGC Domain over interface mId.
GRAR	Provides a network storage capability for state associated to named M2M devices and handles subscriptions to state changes. Key information items include routable addresses or reachability status. The GRAR also acts as a group manager for M2M devices.
GCS	The GCS is the network selection function when the core network is reachable via several alternative access networks or when the gateway owns several routable addresses, according to the service class of the messages to be routed or other policies. It also provides alternative network or communication service selection in case of failures.
GREM	Acts as a management proxy for the NREM.
GSEC	Implements key management, service layer registration, session key management.
GTM	Optional transaction management.

Figure 14.2   Gateway capabilities.

dIa provides registration and authorization primitives for DAs and GAs to the device/gateway, service session management (event reporting or streaming sessions), and read/write/execute/subscribe/notify primitives for objects or groups of objects residing in M2M devices or gateways, as well as group objects managed by the device/gateway capabilities.

– **mId** between an M2M device or M2M gateway and the M2M service capabilities in the network and applications domain. mId provides registration and authorization primitives for DAs and GAs to the M2M core, service session management (event reporting or streaming sessions), and read/write/execute/subscribe/notify primitives for objects or groups of objects residing in M2M devices or gateways, as well as group objects managed by the devices, gateways or in the network core capabilities.

TS 102 921 "machine-to-machine communications (M2M); mIa, dIa and mId interfaces" specifies the actual implementation of these interfaces over several protocol bindings, currently HTTP and CoAP.

14.2.3   Service Capabilities

The service capabilities can reside in the end device (in the acronyms below x=D for device), in the gateway (x=G, Figure 14.2), or in the network (x=N, Figure 14.3). At present, the following M2M service capabilities have been defined:

- application enablement (xAE);
- generic communication (xGC);
- reachability, addressing and repository (xRAR);
- communication selection (xCS);
- remote entity management (xREM);
- security (xSEC);
- history and data retention (xHDR);
- transaction management (xTM);
- compensation broker (xCB);
- telco operator exposure (xTOE);
- interworking proxy (xIP).

Figure 14.4 illustrates the role of service capabilities in the overall ETSI M2M architecture.

NAE	This function registers network applications and is the single contact point for network applications via interface mIa, it hides the service capabilities topology and performs routing towards capabilities.
NGC	Manages the secure transport session establishment and policing according to the messages service class, using NSEC to retrieve session keys.Performs routing between M2M Devices, M2M Gateways, Service Capabilities and M2M Application residing in the Network and Applications Domain. It supports unicast, multicast and anycast.
NRAR	Provides a network storage capability for state associated to named M2M devices, M2M gateways or groups of M2M devices or gateways, and handles subscriptions to state changes. Key information items include routable addresses or reachability status. The NRAR also acts as a group manager for M2M devices and gateways.
NCS	The NCS is the network selection function for devices reachable via several routable addresses, according to the service class of the messages to be routed or other policies. It also provides alternative network or communication service selection in case of failures.
NREM	Provides firmware update support, configuration management functions, performance and fault management. Typically proxies requests to a Broadband forum TR-069 auto-configuration server.
NSEC	Implements key management, service layer registration, session key management. Interfaces with a M2M Authentication server (MAS) (e.g. via Diameter) to obtain authentication data.
NDHR	Optional transaction data archival for legal requirements.
NTM	Optional transaction management.
NCB	Optional financial compensation service.
NTOE	Exposes services such as SMS, MMS, USSD, localization, etc.

Figure 14.3 Network-domain capabilities.

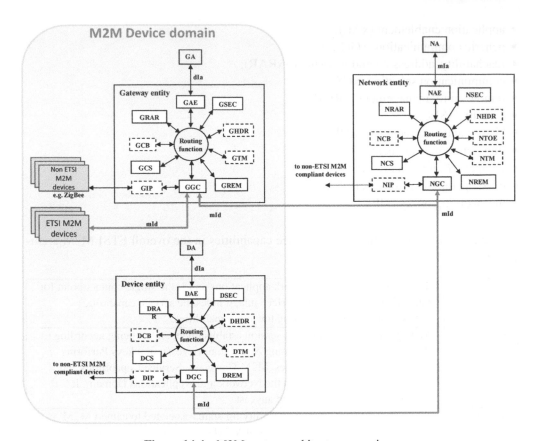

Figure 14.4 M2M system architecture overview.

14.3 ETSI M2M SCL Resource Structure

As expected in a REST architecture style, the functionality of the ETSI M2M service capability layer (SCL) is exposed as a set of addressable resources. Currently, TS 102 921 specifies two serialization options for resources: XML or JSON. For XML, support for FastInfoset (ITU-T X.891 | ISO/IEC 24 824-1) and efficient XML interchange (EXI, http://www.w3.org/TR/exi/) optimizations are recommended.

A given M2M entity exposes its resources as subresources of sclBase. The sclBase resource is represented by an absolute URI. All other resources hosted or registered in the SCL are identified by a URI that is hierarchically derived from the URI of the sclBase. Each subresource "x" of multiplicity 1 is modeled as an attribute named xReference and containing the subresource URI, and each subresource "y" of multiplicity "0..un-bounded" is represented as one collection attribute named yCollection that contains a list of subresource URIs.

The ETSI M2M resource hierarchy is outlined in Figure 14.5. For clarity, not all re-sources currently defined by ETSI M2M are represented. Some subresources are "virtual",

Sclbase	Attributes including references to sub-resources	Sub-resource attributes	
	accessRightID		ID of the access rights resource applicable to this SCL
	creationTime		Creation time attribute
	lastModifiedTime		Last modification time attribute
	pocs		Points of Contact attribute
	searchStrings		Tokens used as keys for searching resources and contents
	sclsReference/		**Reference to SCLs collection subresource**
		accessRightId	Access right id of the parent resource
		creationTime	Creation time
		lastModifiedTime	Last modification time
		mgmtObjReference	Reference to Management object resources
		sclCollection	Reference to registered SCL resources
		subscriptionsReference	Reference to Subscription resources, see Figure 12
	applicationsReference		**Reference to Applications collection subresource**
		accessRightId	Access right id of the parent resource
		creationTime	Creation time
		lastModifiedTime	Last modification time
		applicationCollection	Reference to application resources, see Figure 7
		mgmtObjsReference	Reference to management objects collection resources
		subscriptionsReference	Reference to Subscriptions collection resource
	containersReference		**Reference to Containers collection subresource**
		accessRightId	Access right id of the parent resource
		creationTime	Creation time
		lastModifiedTime	Last modification time
		containerCollection	References to Container resources, see Figure 10
		locationCollection	References to Location resources
		subscriptionsReference	Reference to Subscription collection resource
	groupsReference		**Reference to Groups collection subresource**
		accessRightId	Access right id of the parent resource
		creationTime	Creation time
		lastModifiedTime	Last modification time
		groupCollection	References to group resources, see Figure 11
		subscriptionsReference	Reference to Subscription collection resource
	accessRightsReference		**Reference to AccessRights collection subresource**
		accessRightId	Access right id of the parent resource
		creationTime	Creation time
		lastModifiedTime	Last modification time
		accessRightCollection	Reference to access right resources, see

Figure 14.5 ETSI M2M resource hierarchy overview.

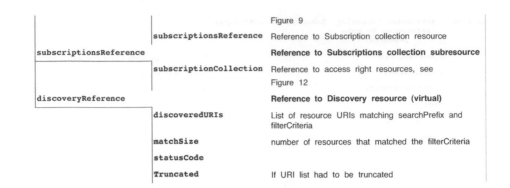

		Figure 9
	subscriptionsReference	Reference to Subscription collection resource
subscriptionsReference		**Reference to Subscriptions collection subresource**
	subscriptionCollection	Reference to access right resources, see Figure 12
discoveryReference		**Reference to Discovery resource (virtual)**
	discoveredURIs	List of resource URIs matching searchPrefix and filterCriteria
	matchSize	number of resources that matched the filterCriteria
	statusCode	
	Truncated	If URI list had to be truncated

Figure 14.5   *(Continued)*

that is, they serve to implement a specific function and typically can only be read and do not have an e-tag.

More detailed views can be found in Figure 14.6 for SCL resources, in Figure 14.7 for application resources, in Figure 14.10 for container resources, in Figure 14.9 for access right resources, in Figure 14.11 for group resources, in Figure 14.12 for subscription resources, in Figure 14.13 for notification channels resources.

Applications register to a SCL simply by creating a new application resource under the applications collection subresource. Remote SCLs register to a local SCL by creating a new SCL resource under the SCLs collection subresource.

Basic REST primitives define how to create, read, update or delete a resource in its entirety, but there are many cases where an application wants to interact with only a part of the resource. TS102921 provides a number of conventions enabling partial resource read, write or subscription. For instance, single attributes or elements can be addressed. For resources formatted as XML content, ETSI M2M supports the XCAP (XML configuration access protocol, RFC 4825) approach, which allows arbitrary xpath expressions to manipulate parts of the XML representation of a resource.

14.3.1   SCL Resources

The sclCollection attribute of the SCLs collection subresource contains references to one or more SCL resources. The structure of an SCL resource representation is outlined in Figure 14.6. represents the specific SCL resource name.

14.3.2   Application Resources

Application resources represent applications (DA, GA or NA). Applications register under one or more SCLs. They interact with the rest of the system via mIa (NA) or dIa (DA, GA)

`<scl>`	Attributes including references to sub-resources	Sub-resource attributes	
	accessRightID		ID of the access rights resource applicable to this SCL
	creationTime		Creation time attribute
	expirationTime		Expiration time attribute
	id		Id attribute
	lastModifiedTime		last modification time attribute
	onlineStatus		ONLINE | OFFLINE | NOT_REACHABLE
	pocs		Points of Contact (network URI) of this SCL
	searchStrings		Tokens used as keys for searching resources and contents
	Schedule		Schedule that tells when an SCL is available
	serverCapability		If TRUE the registered SCL can handle incoming requests.
	discoveryReference		**Reference to Discovery resource** (virtual resource)
		discoveredURIs	List of resource URIs matching searchPrefix and filterCriteria
		matchSize	number of resources that matched the filterCriteria
		statusCode	
		Truncated	If URI list had to be truncated
	notificationChannelReference		**Reference to Notification channel resources**
		creationTime	Creation time of the resource
		lastModifiedTime	Last modification time
		notificationChannelCollection	List of notification channel resources, see Figure 13
	mngtObjsReference		**Reference to Management object collection subresource**
		...	(TBD at the time of writing)
	applicationsReference		**Reference to Application collection subresource**
		(see Figure 5)	
	containersReference		**Reference to Containers collection subresource**
		(see Figure 5)	
	groupsReference		**Reference to Groups collection subresource**
		(see Figure 5)	
	accessRightsReference		**Reference to AccessRights collection subresource**
		(see Figure 5)	
	subscriptionsReference		**Reference to Subscriptions collection subresource**
		(see Figure 5)	

Figure 14.6   SCL resource representation.

Figure 14.7   Detail of a DA/GA/NA representation.

```
GET /m2m/applications/firstApp HTTP/1.1
Host: m2m.example.com

HTTP/1.1 200 OK
Content-Type: application/xml

<?xml version="1.0" encoding="UTF-8"?>
<application xmlns="http://uri.etsi.org/m2m">
    <creationTime>2001-12-31T12:00:00.000</creationTime>
    <searchStrings>
            <searchString>tag1</searchString>
            <searchString>tag2</searchString>
    </searchStrings>
    <containersReference>
            http://m2m.operator.org/m2m/applications/firstApp/
containers
    </containersReference>
</application>
```

Figure 14.8   Example application resource application/xml serialization.

interfaces. Basic interactions may involve reading or writing any resource accessible by the SCL (typically reading or writing container subresources), provided that the application has the relevant access rights to the resource (see Figure 14.9). This notion of generic access right is one of the key features of ETSI M2M, enabling secure sharing of the M2M network, gateways and devices by multiple applications.

Figure 14.8 shows an example serialization of an application resource, using application/xml.

<Access right>	
announceTo	list of resources to which the resource is announced at the moment
creationTime	Creation time of the resource
expirationTime	
lastModifiedTime	Last modification time
searchStrings	Tokens used as keys for searching resources and contents
permissions	Permissions
selfPermissions	Permissions on this resource
subscriptionsReference	Reference to Subscriptions collection subresource
	(see Figure 5)

Figure 14.9   Access right resource representation.

This type of interaction based on reading or writing resources is adapted to many situations, but not to all. For instance, an application may have a transient variable that it wishes to expose to other applications (for instance a timer), but it would clearly be inefficient to report the value of that transient variable to a network buffer. Other applications may offer access to a very large dataset, and the probability of external applications requesting access to some items in the dataset is very low, therefore publishing the whole dataset would be a very inefficient model.

The notion of application point of contact (aPoC) provides a more dynamic interaction model with the application: retargeting. This feature requires that the application register an application point of contact (see "aPoC" attribute in the resource tree). When an application does not provide an application point of contact, the SCL retargeting behavior is not available. In addition, the application may publish a set of application point of contact paths ("aPoCPaths" attribute), in order to restrict the SCL retargeting behavior to only resources addressed by specific subpaths. In the absence of this parameter, the application accepts that any resource request may be retargeted if it has published an application point of contact. Each application point of contact path is a tuple containing:

- A path relative to the URI of the application path resource;
- An optional access right identifier., which applies to all subresources subordinate URIs of under this path. In the absence of such access right identifier, the access rights of the application registration resource apply.

The REST requests of an external application trying to interact with a subresource identified by a URI matching any of the application point of contact paths prefixes is retargeted by the SCL to the aPoC of the application. This enables the application to respond to the request directly without a need for an intermediary buffer.

14.3.3   Access Right Resources

The access right resource (Figure 14.9) is used to configure the permitted actions for a given resource according to the action issuer.

The permissions attribute contains a list of "permission" elements. Each permission element is a triplet that associates:

- a permission id;
- permission flags (READ, WRITE, DELETE, CREATE and DISCOVER);
- permission holders (all, or specific permission holder URIs and specific domains identified by a path prefix).

14.3.4   Container Resources

Container resources provide a generic buffering capability which facilitates the exchange of data between applications, particularly for applications residing on sleeping devices.

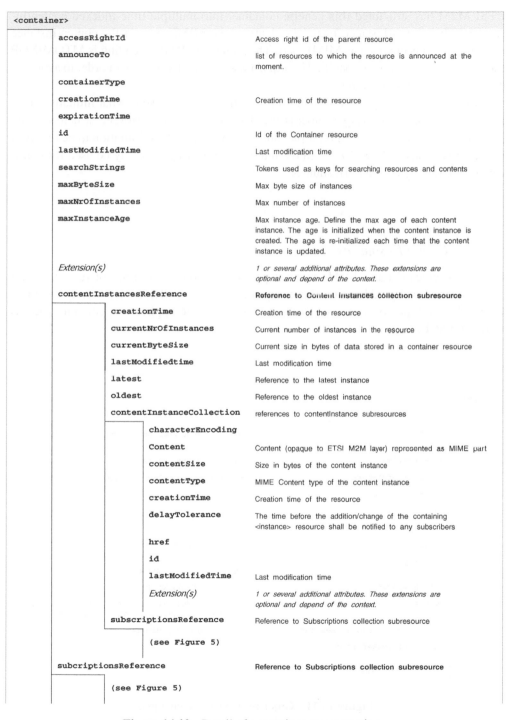

Figure 14.10   Detail of a container representation.

ETSI M2M has structured this generic container into multiple time-indexed contentInstances. The contentInstance metadata identifies the MIME type of the contentInstance content, which is encoded as MIME part. The preferred MIME encoding is MTOM/XOP (application/xop+xml) as specified in http://www.w3.org/TR/xop10/ in order to avoid the overhead of base64 encoding.

This choice of a single index (time) for content instances is somewhat limitative compared to advanced metering protocols that handle multi-index storage structures (and provide read queries based on multiple indexes). We expect this limitation to be removed in future versions of ETSI M2M, in the interim searchstring tags may be used for multi-category indexing of contentInstances.

14.3.5   Group Resources

Group resources (Figure 14.11) make it possible to address a group of resources in a single operation, by addressing the group membersContentReference virtual resource. Currently, the types of resources that can be grouped are application, container, access right, and SCL.

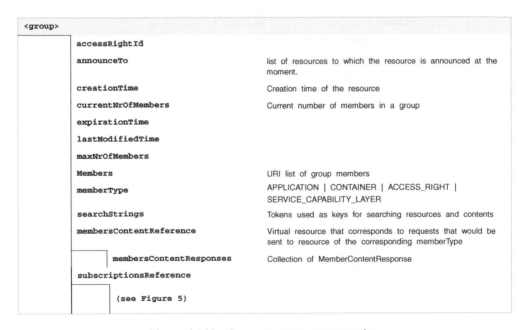

Figure 14.11   Group resource representation.

<subscription>	
contact	URI where the subscriber wants to receive its notifications
creationTime	Creation time
delayTolerance	The slack time allowed to notify a resource change
expirationTime	Expiration time for this subscription
filterCriteria	Optional reference e-tag and lastModifiedTime value of the subscribed resource. A notification should be immediately triggered if the resource is newer.
lastModifiedTime	
minimalTimeBetweenNotifications	Minimal time between notifications in milliseconds.
subscriptionType	ASYNCHRONOUS I SYNCHRONOUS

Figure 14.12   Detail of subscription representation.

14.3.6 Subscription and Notification Channel Resources

Subscription resources provide RESTful support for the subscribe/notify primitives that are made available to applications at the functional level.

Clients willing to subscribe to changes of a given resource must add a subscription subresource (Figure 14.12) under the subscriptions collection. The notification can happen asynchronously (subscriptionType=ASYNCHRONOUS) in which case the client must specify a contact URI as part of the subscription) or may use the "long-polling model". The long-polling model is used by applications that are not server capable. Such applications must first request the SCL to create a notificationChannel resource (see Figure 14.13). The SCL will populate the newly created notificationChannel with two parameters: the contactURI to be used in subscriptions, and the associated long polling URI in the channelData attribute. The nonserver-capable application will then read the long-polling URI, and the SCL will reply only when notifications happen related to the associated contactURI.

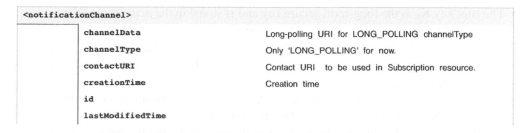

<notificationChannel>	
channelData	Long-polling URI for LONG_POLLING channelType
channelType	Only 'LONG_POLLING' for now.
contactURI	Contact URI to be used in Subscription resource.
creationTime	Creation time
id	
lastModifiedTime	

Figure 14.13   Detail of notification channel representation.

The notification content includes the subscription URI, as well as the latest representation of the subscribed-to resource.

14.4   ETSI M2M Interactions Overview

An application may interact with resources on the local SCL provided that the accessRights of that resource allow the type of action for the issuer application. An application may also interact with resources located on other SCLs, in which case the local SCL forwards the request to the target SCL, and access rights are checked by the target SCL.

ETSI M2M provides several options for the handling of request responses:

- **Synchronous response:** the expected response content is delivered in the response of the original request, which is blocked until a response becomes available or a timeout error occurs.
- **Semiasynchronous response:** requests include correlation data. If the expected response content cannot be delivered immediately, the target SCL sends an acknowledge indication and an indication of the minimum time to wait before sending a new request. The issuer SCL is expected to reissue the request with the same correlation data, until it obtains the expected response content in the target SCL response.
- **Asynchronous:** the local SCL needs to be server capable, and provides a contact_server URI for responses, as well as correlation data in each request, so that responses received over the server URI can be associated to the corresponding requests.

14.5   Security in the ETSI M2M Framework

14.5.1   Key Management

ETSI M2M uses a key hierarchy:

- The root key K_R is the long-term secure key and is stored in the network and the device. It is used to derive the service keys. Depending on the underlying network technology K_R can be stored at the network level in a secure element (e.g., SIM card) or at the service capability layer. K_R is typically preinstalled and shared between the network and the gateway.
- The service key K_S is derived from the root key during the authentication and authorization of the M2M service layer. K_S is derived during the gateway-registration procedure to the network, and associated to the session (it may expire and be renewed during the session). The service key is not available to applications.

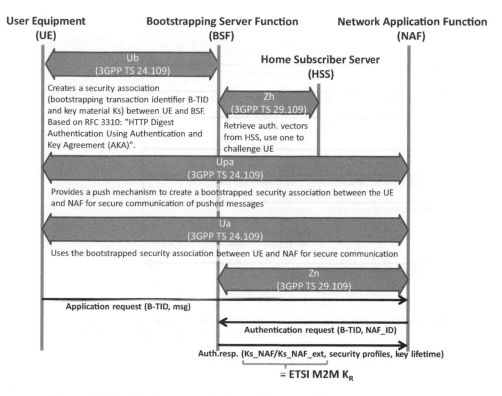

Figure 14.14 Generic authentication architecture (GAA, 3GPP TS 33.220).

- The optional application key K_A. There is one application key per application that wishes to manage its own security context. K_A will be used for authentication and authorization of M2M applications at the M2M device or gateway. At the time of writing, the exact use of K_A was not specified.

TR102921 describes the several security bootstrapping mechanisms. Some options leverage on existing security bootstrapping frameworks at the access-network level, for example, the 3GPP generic bootstrapping architecture (GBA over the 3GPP Ub interface defined in 3GPP TS 24.109, see Figure 14.14) that uses SIM or USIM card security material to derive a shared key between the M2M gateway (user equipment) and the MSBF (acting as 3GPP NAF, via the NSCL).

Other options are completely independent of the underlying transport network, for example, using EAP over PANA transport, with the M2M gateway acting as EAP peer and PANA client, the NSCL as EAP authenticator and PAN authentication agent, and the MSBF acting as EAP authentication server (Figure 14.15).

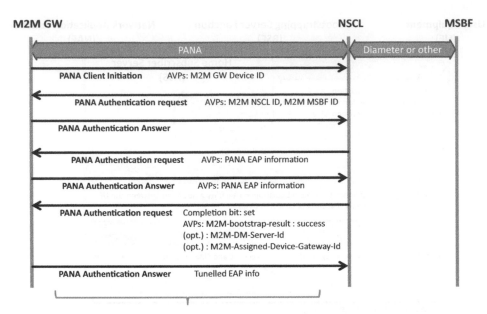

K_R = Hash(EAP Extended Master Session Key, "ETSI M2M Device-Network Root Key" |
M2M-Assigned-Device-Gateway-Id | M2M-Domain-Id)

Figure 14.15   PANA/EAP ETSI M2M security bootstrapping.

14.5.2   Access Lists

The ETSI M2M framework is designed to enable usage of a shared M2M infrastructure
(network, gateways, sensors and devices) by multiple applications.

The security aspects are very important in this context:

– A given end-user (e.g., a residential subscriber) owning several sensors under control
of his local M2M gateway will of course not want to publish sensor data, or provide
access to actuators, to anyone. ETSI M2M provides an access-control list (ACL) for
most resources published by the M2M gateway SCL. In the example of Figure 14.7,
the user, by configuring properly the ETSI M2M gateway SCL ACLs can decide which
network application can perform which REST action (e.g., read or write) on which
device application, or even on which container of which device application.

Of course, the average end user will never deal with the details of ETSI M2M. Instead,
a user-friendly interface will be provided by the M2M service provider. In the example
of Figure 14.16, the M2M service provider presents each "network application" (or
device application, or group of related NA and DA applications) as an icon, using the
now-familiar application store design. In Figure 14.17, the user is configuring a sensor,
and is asked which applications can access this sensor information: behind the scenes,

Figure 14.16 A possible user interface for ETSI M2M service configuration.

the middleware will configure the proper ACLs in the representation of that sensor on the gateway SCL.

14.6   Interworking with Machine Area Networks

ETSI M2M provides a very interesting model to interact with multiple machine area networks in a homogeneous way, enabling network or gateway applications to interact with these networks with minimal customization requirements. For most applications, only the specific sensor/actuator URIs and parameter names need to be configured or discovered: the actual underlying protocol details of the machine area networks are abstracted by the ETSI M2M framework.

ETSI TR 102 966 "Interworking with M2M Area Networks" explores the specific mappings recommended for interworking with other M2M technologies, such as ZigBee.

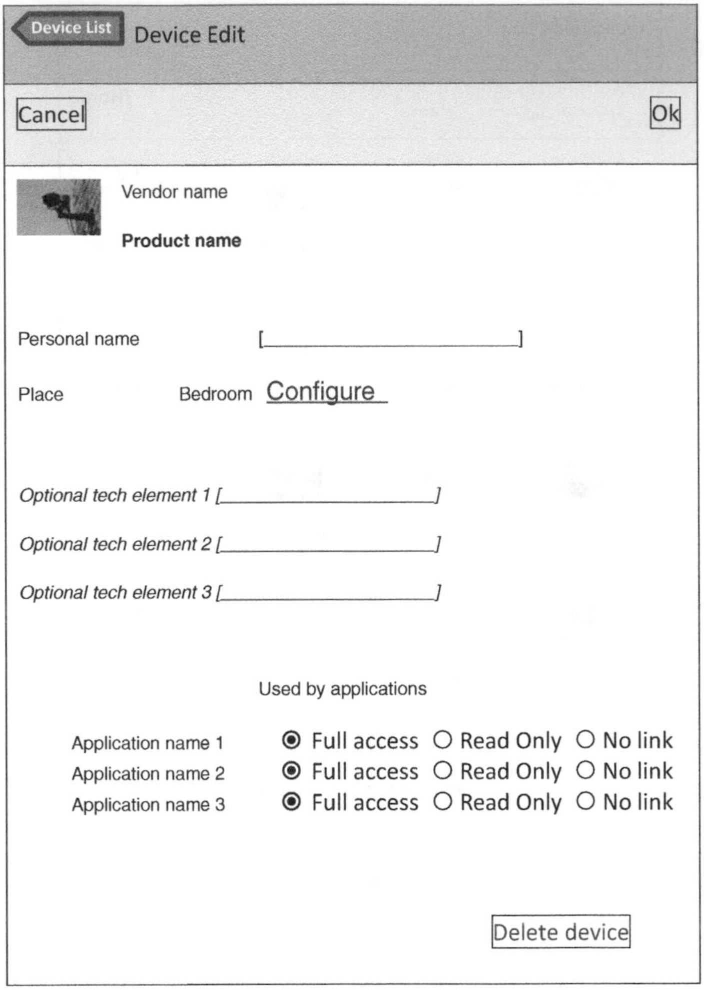

Figure 14.17 Security configuration screen for a sensor.

At the time of writing this section, the document was still in early draft stage, however, the author was an active participant in the elaboration of this technical report, and believes that the approach outlined below will be very close to the final recommendation.

14.6.1 Mapping M2M Networks to ETSI M2M Resources

There are many potential ways of mapping M2M networks to ETSI M2M standard resources. One possibility is to attempt to recreate the hierarchy of native M2M networks within the hierarchy of ETSI M2M resources (e.g., SCLs for top-level networks, applications for nodes, etc. . . .). However, obviously, it is impossible to replicate the exact resource hierarchy of any network, which can get quite deep. In the case of ZigBee for instance (refer to Chapter 7), each controller may access a collection of networks, each with a number of devices hosting a number of applications. Each application publishes an interface composed of a number of clusters, and each cluster is composed of multiple attributes and primitives!

The preferred option is to represent external machine area networks as a collection of applications (GA or DA). Any resource hierarchy present in the native M2M network is recreated by means of pointer attributes: the ETSI M2M application representing the parent object will expose special attributes pointing to child objects.

Recognizing that most automation protocols use a similar datamodel structure, TR 102 966 introduces specific 'tags' (special searchString values) to identify the type of 'object' modeled by the application in a protocol independent way:

- ETSI.ObjectType tags (for instance ETSI.ObjectType/ETSI.AN_NWK) are used to discriminate between applications representing Interworking Proxy objects (ETSI.ObjectType/ETSI.IP), Network objects (ETSI.ObjectType/ETSI.AN_NWK), Device objects (ETSI.ObjectType/ETSI.AN_DEV), Application objects (ETSI .ObjectType/ETSI.AN_APP), and Point objects (ETSI.ObjectType/ETSI.AN_POINT). The various object types will be illustrated in the example of ZigBee (see Section 14.6.2).
- ETSI.ObjectSemantic tag (for instance ETSI.ObjectSemantic /OASIS.OBIX_1_1) is used to discover the semantic conventions supported by the object. The syntax of an object representation is usually indicated by its Content-Type, for instance application/xml. However, multiple semantic conventions may leverage the same syntactic rules. In the use case of interworking with control and sensor networks, examples of such semantic conventions leveraging application/xml syntax are OASIS oBix (ETSI.ObjectSemantic/OASIS.OBIX_1_1), ZigBee Gateway Device REST binding, or ASHRAE BACnet annex am (ETSI.ObjectSemantic/ASHRAE.CSML_1_0).

  ETSI M2M implements a REST design model that allows multiple representations of the objects manipulated through the ETSI M2M SCL.
- ETSI.ApplicationProfile tags are reserved for future use. The intent is to be able to facilitate search of specific devices for example, "lamps". Nomenclatures have been created by ZigBee, KNX and LonWorks, and one is being worked on by BACnet. Future work would lead to a harmonized nomenclature that would use this tag category

14.6.2   Interworking with ZigBee 1.0

Interworking with ZigBee 1.0 networks leverages the ZigBee Alliance "Gateway specification for network devices", version 1.0 (refer to Chapter 7 for details). This specification exposes how a ZigBee gateway devices (ZGD) exposes the ZigBee network resources to IP host applications (IPHA), over several network bindings.

An ETSI M2M application (local gateway application or device application in the case of a separate ZigBee interworking device) models the ZGD and acts as an

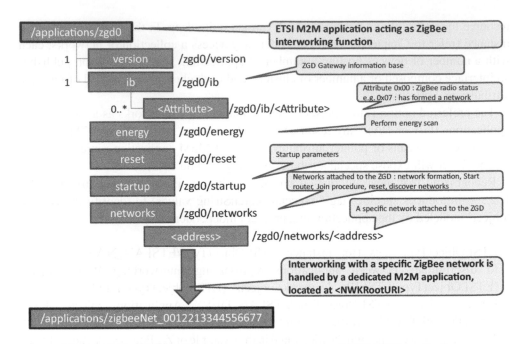

Figure 14.18   ZGD REST representation as a standard ETSI M2M application.

interworking proxy. It has a searchString attribute containing technology-independent tag "ETSI.ObjectType/ETSI.IP" so that applications may discover this application, which has interworking proxy capability. This application contains one <container> subre- source that has a searchString containing tag "ETSI.ObjectType/ETSI.IP", and a tag of category ETSI.ObjectSemantic (oBix for our examples below). The latest contentInstance of this container lists the networks supported by this interworking proxy (in a protocol- independent way) as well as proxy capability information. Each network element points to the ETSI M2M application that represents it. This container instance may also in- clude links to the ZGD representation defined by the ZigBee alliance (Figure 14.18). An example interworking proxy representation is shown in Figure 14.19.

Each ZigBee network is represented by a specific ETSI M2M application located at the URI specified in the 'networks' list. Applications representing a network have a searchString containing tag: ETSI.ObjectType/ETSI.AN_NWK, and one container with the same tag. The latest contentInstance of this container contains the representation of the network. Figure 14.20 illustrates how a client application can read this resource.

The representation of the network contains the network identifier and the list of nodes (in a protocol independent way), as well as technology-specific information related to the network. Each ZigBee node is represented by a specific ETSI M2M application located at the URI specified in the 'nodes' list. Applications representing a node have a searchString containing tag: ETSI.ObjectType/ETSI.AN_DEV, and one container with the same tag.

```
>>> HTTP GET
/gsc/applications/zgd0/containers/descriptor/contentInstances/
last/content

<<< 200 OK
<obj>
     <str name="interworkingProxyID" val="Text for correla-
tion purpose"/>
     <list name="supportedTechnologies">
          <obj>
                    <enum name="anStandard" val="ZigBee_1_0"/>
                    <enum name="anProfile" val="ZigBee_HA"/>
                    <enum name="anPhysical"
val="IEEE_802_15_4_2003_2_4GHz"/>
          </obj>
     </list>

     <list name="networks"/>
                    <ref href="/gsc/applications/zbnw0/">
     </list>
</obj>
```

Figure 14.19   Retrieving the M2M application resource modeling the ZGD via HTTP GET.

```
>>> HTTP GET
/gsc/applications/zbnw0/containers/descriptor/contentInstances/
last/content

<<< 200 OK
<obj>
     <str name="networkID" val="Text for correlation
purpose"/>
     <str name="extendedPanID" val"0x685B3C34"/>

     <list name="nodes">
               <ref href="/gsc/applications/zbnode0/"/>
     </list>
</obj>
```

Figure 14.20   Retrieving the M2M application resource modeling the ZigBee network via HTTP GET.

```
>>> HTTP GET
/gsc/applications/zbnode0/containers/descriptor/contentInstances/
last/content

<<< 200 OK
<obj>
        <str name="nodeID" val="Text for correlation purpose"/>
        <str name="ieeeAddress" val="0x685B3C88"/>
        <enum name="type" val="endDevice"/>

        <list name="applications">
                <ref href="/gsc/applications/zbapp0/"/>
        </list>
</obj>
```

Figure 14.21 Retrieving the M2M resource for a node via HTTP GET.

The latest contentInstance of this container contains the representation of the node. Figure 14.21 illustrates how a client application can read this resource.

Our example network has a single node. Each node may host multiple applications, for instance multiple switches or lamp controllers. ZigBee identifies each application as an 'endpoint'. Each ZigBee endpoint is represented by a specific ETSI M2M application located at the URI specified in the node 'applications' list. ETSI M2M Applications representing a node application[1] have a searchString containing tag: ETSI.ObjectType/ETSI.AN_APP, and one container with the same tag. The latest contentInstance of this container contains the representation of the node application, in this case a ZigBee endpoint. Figure 14.22 gives an example representation of a node application resource.

Each node application may implement multiple interfaces. ZigBee identifies each interface as a 'cluster'. Interfaces may be represented as separate applications (the element is a reference to the M2M application), but in the example of Figure 14.22 the contentInstance data includes the actual interface content. Each interface contains protocol dependent primitives (e.g. 'toggle') as well as a number of attributes (e.g. 'OnOff'). TR 102 966 offers an abstraction for attributes representing measurements: the Point that may include protocol-independent specification of units, range, etc. (It is not used in the example).

When the ZigBee interworking application starts, it audits all networks and replicates a view of the ZigBee network topology in the GSC (Figure 14.23). This view includes all ZigBee nodes and all ZigBee applications (endpoints). When a new ZigBee node is

---

[1] Node applications are applications running in sensors, not to be confused with the M2M application resources that model them.

```
...
<int name="endpoint" val="1"/>
<int name="applicationProfileID" val="0x0104"/>
<int name="applicationDeviceID" val="0x0100"/>

<list name="Interfaces">
     <obj>
            <str name="clusterID" val="0x0006"/>
            <enum name="clusterType" val="input"/>

            <list name="attributes">
                <ref name="0x0000"


    href="/<sclBase>/applications/<networkX_nodeY_application
Z>/containers/0x0006_OnOff"/>
            </list>

            <list name="operations">
                <op name="0x00"
href="/<sclBase>/applications/<interworking_proxy_unit>/0x0006_
off"/>
                <op name="0x01"
href="/<sclBase>/applications/<interworking_proxy_unit>/0x0006_
on"/>
                <op name="0x02"
href="/<sclBase>/applications/<interworking_proxy_unit>/0x0006_
toggle"/>
            </list>
     </obj>
</list>
...
```

Figure 14.22   Extract of the ZigBee endpoint representation.

added in the network, the ZigBee HAN representation is immediately updated to reflect this change in the GSC.

At this stage, other gateway or network applications can discover the elements of the ZigBee networks by retrieving the resources maintained in the GSC by the driver. They can also subscribe to the representation changes, if the driver supports it (and creates the appropriate subscriptions collections in the ZigBee network representation).

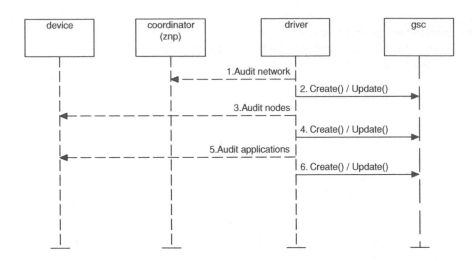

Figure 14.23 ZigBee driver explores the ZigBee network and creates the SCL representations.

14.6.3 Interworking with C.12*

In order to prove the versatility of ETSI M2M standards and their ability to provide a common framework for interfaces to existing standards, we explore in this section the interface to C.12 metering standards. As mentioned in Chapter 10, the communication protocol for the transport of C12.19 tables requires only basic read and write services. Such standardization of communication patterns using few verbs is typical of REST design principles.

As ETSI M2M decided to use a RESTful approach for specifying its interface between M2M applications and the service capability layer, we attempt to use this interface to interact with C12.19 tables accessible over a C12 network. At first sight, a REST-style interface can be achieved by mapping each C12.19 table to a corresponding REST resource. The basic read and write services can also easily be mapped on the classical CRUD verbs (create, retrieve, update, delete).

In the DLMS/COSEM case (see Chapter 11), the data communication services accessible to the metering application are a little more complicated than read/write due to the existence of the "action" service. An intermediate resource has to be used in order to communicate the name of the "action" and the associated parameters in a REST style. In the C12.19 data model it is easier since the possible "actions" are already included in two specific tables (Tables #07 and #08). The "action" is called "procedure" and a request for executing a specific procedure corresponds to the invoke of the basic write service for Table #07.

Our purpose here is not to propose a specification of what would interface to native RESTful C12 meters, but instead we attempt to define an interworking function between

---

*This section is contributed by Jean-Marc Ballot.

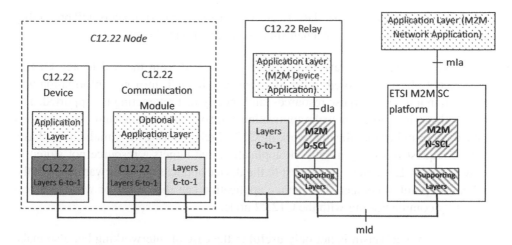

Figure 14.24 Interworking function implemented as C12.22 relay.

a standard C12 network and ETSI TC M2M, that is, exposing an ETSI M2M application interface. As an example, implementation for such interworking, we consider a C12.22 relay, as represented in Figure 14.24.

In Figure 14.24, mIa, dIa and mId are standardized ETSI M2M reference points. N-SCL and D-SCL are the ETSI M2M "network service capability layer" and "device service capability layer".

On the left side of the C12.22 relay, the communications are fully compliant with the ANSI C12 suite of standards. On the right side of the C12.22 relay, the dIa, mId and mIa reference points support the RESTful communication patterns.

From the network application perspective, the C12.22 node can be represented as part of the C12.22 relay "device application" representation, or each node could be represented as a separate device application. During initialization steps, the device application in the C12.22 relay can read the C12.19 table #00 (called GEN_CONFIG_TBL) of the C12.22 node in order to get the full list of the supported tables and procedures. Then, the C12.22 relay is able to use the RESTful dIa interface in order to request the creation of a set of REST resources that represent the set of C12.19 tables handled by the C12.22 node. A one-to-one mapping is performed between C12.19 tables and REST resources. The resources are physically created in the N-SCL (they could also be created locally in the C12.22 relay if the relay has REST server capability).

The interactions with these resources can be performed by using the CRUD verbs enhanced with notify and subscribe as specified by ETSI TC M2M. When a network application wants to read (or write) a C12.19 table contained in the C12.22 node, it uses a retrieve (or an update) query to the corresponding REST resource. The C12.22 interworking relay then contacts the relevant C12.22 node in order to read/write the actual content of the relevant C12.19 table. The C12.22 interworking relay may represent static

C12 node parameters as container data. Read/writes to dynamic parameters will preferably use the ETSI M2M retargeting mechanism already described above:

- The REST resources that have to be directly requested from the real legacy C12.22 node are not physically stored in the SCL, instead the SCL mirrored information will only contain a pointer to these resources, and specify that retargeting is supported. The retargeted queries can be sent to the application point of contact, which must be published by the device application (the interworking C12.22 relay) during the initialization phase.
- When a request is sent by a network application that targets a resource not stored in the SCL, the SCL retargets the request to the device application by forwarding it to the application point of contact. The device application can then read (or write) the real content by communicating with the C12.22 node.

This redirection mechanism is not only useful in the case of interworking but also make it possible to implement natively ETSI M2M compliant C12 meters. Such a C12 meter would implement a RESTful dIa interface, and would be able to mirror static C12.19 REST resources in the SCL. For C12.19 tables that are volatile or too frequently changing (e.g., time elapsed since a specific event, time remaining before a specific event, consumption metering counters), the redirection mechanism is used instead and the mirrored representation would store only pointers.

14.6.4   Interworking with DLMS/COSEM

ETSI TC M2M decided to use a RESTful approach for the interface between M2M applications and the service capability layer. The interworking with DLMS/COSEM meters is not trivial because DLMS/COSEM uses an object oriented approach that is not RESTful compliant.

DLMS/COSEM meters are "legacy devices" according to the ETSI M2M reference architecture document and interworking will be performed by an interworking proxy (NIP or GIP), located in a M2M gateway (GIP) or in the core network (NIP). At the time of writing, ETSI TC M2M had not yet fully specified the interworking proxy function. In the following text, we are proposing a possible approach in line with the current specification, based on a network-based interworking proxy (NIP), and illustrated on Figure 14.25.

We are assuming that the network application interacting with the DLMS meter is ETSI M2M compliant. The NIP could be implemented as a new interface on DLMS data concentrators. This would be an option to standardize the interactions between network applications and data concentrators (currently out of scope of DLMS/COSEM).

The network interworking proxy (NIP) is considered as a part of the ETSI M2M platform. It contains a COSEM client AP and a COSEM client AL. The communications between the client parts in the NIP and the server parts in the metering equipment use the DSML/COSEM specifications. Once the client AP has established an association with the server AP in the meter, a set of the COSEM interface objects becomes accessible.

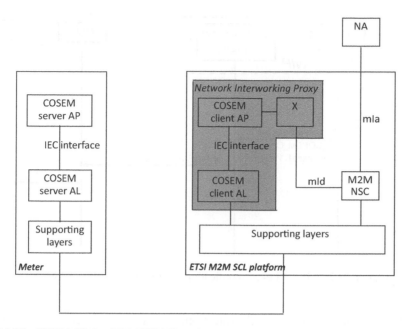

Figure 14.25 ETSI M2M – DLMS/COSEM interworking using a network interworking proxy (NIP).

The "X" part of the NIP in Figure 14.25 represents the new software added for ETSI M2M interworking.

From a RESTful point of view, each COSEM interface object can be represented as a resource: the NIP first reads each COSEM interface object by using the standardized COSEM GET service. Then, after reading the object, the corresponding resource is created in the SCL layer. Only static parameters should be stored in the SCL. Volatile parameters such as counter values should be listed in SCL resources only as indirections, that is, read requests from network applications will be retargeted by the SCL to the interworking proxy.

The DLMS specifications also define an ACTION service enabling a client AP to remotely invoke one or more methods of COSEM interface objects. This interaction pattern can also leverage the ETSI M2M retargeting mechanisms: SCL representations of COSEM interface objects will list a subresource for each supported action. The NA may invoke an action by sending a POST request targeted to the corresponding subresource (and containing the action parameters). The write request will be retargeted to the interworking proxy by the SCL, and the interworking proxy will convert the request to a native DLMS/COSEM action service invocation. Synchronous responses will be provided in the 200 OK response.[2]

---

[2] In the case of asynchronous responses, the NIP response may just be 201 created, and the asynchronous response may be sent later to a resource supplied a part of the invoke parameters.

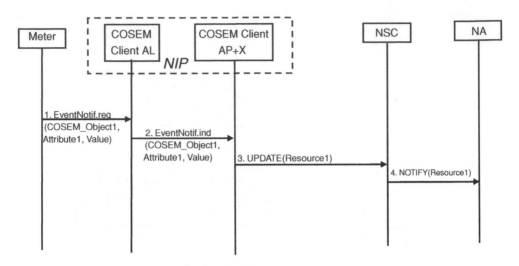

Figure 14.26   DLMS event-notification interworking with ETSI M2M.

The meter may also send an EventNotification, this case represents the exception of the DLMS client–server model. This notification contains the value of a COSEM interface object attribute. This interworking case is illustrated in Figure 14.26.

In step 1, the meter needs to inform the client AP of the value of one attribute (Attribute1) of one COSEM object (Object1). The meter uses the COSEM EventNotification service in order to provide the client AP with the attribute value (step 2).

The NIP has to update Resource1 that corresponds to Object1 in the REST environment. As a single attribute of Object1 is modified, a partial update is performed.[3] The network application will be notified of this update if it subscribed to Resource1.

14.7   Conclusion on ETSI M2M

Just like networked information systems played a fundamental role in the transformation of almost every business, connected objects will fundamentally change the design of most industrial and automation processes (see for instance draft-ietf-6LoWPAN-usecases). The Internet emerged as the information backbone interconnecting all information systems, and the Internet of Things is now emerging as the backbone interconnecting all objects.

A common view is to consider that the IoT will be just one giant peer to peer IPv6 network (there are about 6×10^{27} IPv6 addresses per square meter on earth). But there are several problems with this view:

---

[3] However, if partial updates are not allowed, due to some internal policy in the REST server, the NIP will have to retrieve Resource1, locally update its content with the new value of the Attribute1, then perform a full update of Resource1.

- For most actual sensor networks using low-bitrate and lossy networks (LLNs), standard IPv6 is not usable: 6LoWPAN will be used instead. The usable addressing space, notably for destination addresses, is very limited and not optimized for any to any worldwide communication. 6LoWPAN is optimized for a local or regional sensor network.
- Basic security considerations force the introduction of application-level gateways between private domains and the public side of the internetwork. Today, the much-touted exhaustion of IPv4 addresses does not have the screaming halt effect one might expect on the development of the Internet: this is because all corporations use private addresses internally, and their firewalls map on demand these private addresses to very few external public IP addresses using NAPT (network address and port translation) or application-level proxies, e.g., for HTTP). The same will probably go for the Internet of Things, private 6LoWPAN clouds will be interconnected by application-level gateways.

ETSI M2M is currently the only candidate standard for such "inter-networks of things" application-level gateways. It provides a standard data model for machine area networks, a standard access control list format, and a global application-level addressing mechanism for connected objects. This design ensures that any application on earth can access any sensor, but not using the sensor IPv6 address directly (it may be a private address), using instead the application level URI for the sensor that can be used to route the application requests to the proper object network access gateway, under control of its ACLs. Such gateways are also the ideal place for fieldbus protocol interworking, based on the normalized REST representation of each fieldbus network.

We believe the ETSI M2M framework will become the necessary complement of 6LoW-PAN for service providers envisioning very large scale, multipurpose M2M infrastructure deployments.

Part Five

Key Applications of the Internet of Things

15

The Smart Grid

15.1 Introduction

The smart grid will be one of the most important applications of the Internet of Things.

A major paradigm change is happening in electricity markets, driven by the convergence of several factors:

- The challenges posed by the accelerated introduction of renewable-energy sources in the overall electricity production, which brings an increasing degree of randomness to the traditionally deterministic supply side.
- The ubiquitous penetration of the Internet in homes and businesses, and the increased confidence in next-generation smart distributed networks for mission-critical applications (after years of experience of the successful migration of telephony networks to VoIP).
- The gradual opening of electricity markets, with new regulation opening production facilities and distribution networks to all actors, greater fluidity in electricity trade markets, and fast maturing of the regulatory framework for active utility operators, such as those implementing demand response.
- The increasing volatility of electricity prices, resulting from the underlying volatility of oil and natural gas, but also increasingly from the propagation of external shocks, such as exceptional climatic events, through energy exchanges.

The current credo of electricity operators "demand is unpredictable, and our expertise is to adapt production to demand", is about to be reversed into "production is unpredictable, and our expertise is to adapt demand to production".

As the rules of the game change, the key assets of an energy operator will no longer be the means of production, but the next-generation communication network and information system, which they still need to build entirely. M2M communications and emerging standards such as 6LoWPAN, RPL and ETSI M2M will be key enablers for this evolution.

The Internet of Things: Key Applications and Protocols, First Edition.
Olivier Hersent, David Boswarthick and Omar Elloumi.
© 2012 John Wiley & Sons, Ltd. Published 2012 by John Wiley & Sons, Ltd.

15.2   The Marginal Cost of Electricity: Base and Peak Production

As for any other production type, the cost of electricity generated by a power plant is the sum of the cost of the primary energy supply and the amortization of the plant itself. For this reason, electricity produced by plants running continuously (amortized over the full year) is always cheaper than electricity produced by plants running only sporadically. Nuclear power plants, which produce (relatively) cheap electricity, typically operate continuously except during planned maintenance periods and their production can be adjusted only marginally, over long periods of time (48 h or more).

As a result, demand prediction is key for all electricity operators, as it allows them to plan their production investments. Demand can be decomposed into:

- "Base demand": this is the component of demand that varies most slowly and can be produced from plants running continuously close to their maximum capacity. Electricity production for the base demand has the cheapest marginal production cost.
- Variable demand, which can be supplied from plants operating at a lower utilization rate, or purchased on the market. It has a higher marginal production cost. The "daily peak demand" represents the levels of demand reached only a few hours per day, typically at 7 p.m. in the evening. The yearly peak demand represents the levels reached only a few hours per year, and consequently at the highest marginal cost.

For all electricity operators the marginal production cost gets higher and higher as the current production level increases beyond the "base demand" (Figure 15.1).

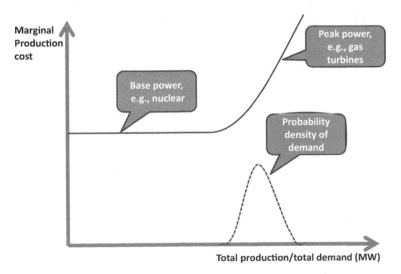

Figure 15.1   The marginal cost of power as a function of current demand/production level.

At some point, the cost of generating each additional MWh gets so high that it may exceed the final, usually fixed or slowly varying, selling price of the MWh to the end consumer. At this point, demand response becomes a no-brainer: instead of producing more power and losing money, the operator should create an incentive for its customers to use less energy.

15.3   Managing Demand: The Next Challenge of Electricity Operators . . . and Why M2M Will Become a Key Technology

Today, most operators model their residential and small-business customers statistically, using a generic demand profile that depends only on the average yearly consumption of the customer and his location. Operators only measure the total consumption of each customer, not its distribution over time. As a result, an operator's ability to reward their customers that have a lower "peak demand" component in their consumption is limited to simple time-based tariffs (typically a day and a night tariff for counters able to measure day and night consumption separately).

With more advanced counters able to report consumption profiles over time (such as those deployed in first-generation automatic metering projects), operators can introduce time-dependent tariffs that give customers an incentive to minimize their peak consumption. If such tariffs are published and are reasonably stable, users can schedule their consumption (e.g., use delayed start function of washing machines or dryers) and therefore "flatten" their consumption pattern and reduce their average electricity bill.

But unfortunately, published time-dependent tariffs are not sufficient: the increasing proportion of production coming from renewable sources creates random, unpredictable changes in total production power. The randomness of production power requires real-time adaptation of demand: tariffs would need to become dynamic, that is, change in real time, or if tariffs remain published, separate real-time incentives to adapt demand must be created.

Obviously, most residential customers or small businesses do not have the time or expertise required to control their heating, air conditioning, hot water tank, fridge, or ventilation system in real time. This is why commercial electricity suppliers will also need to actively manage the regulation of energy use in each home, using bidirectional M2M technology. "Intelligent meters" will be useless if not complemented by active energy management systems. Such bidirectional M2M energy-management systems will flatten the demand profile, and will also be able to react in real-time to network or production incidents, minimizing the costs associated to last-minute energy purchases.

In the future, the ability to provide such efficient demand response programs will become a key differentiator for electricity operators: the operators who will manage to best "flatten" the consumption of their customers, while preserving or enhancing comfort, will reduce their production costs and become more competitive.

15.4   Demand Response for Transmission System Operators (TSO)

15.4.1   Grid-Balancing Authorities: The TSOs

While multiple commercial electricity suppliers may exist in a country, they all share the same distribution grid. Because the network is interconnected, any commercial supplier can buy additional capacity from independent producers or other utilities and easily distribute this power to its own customers.

But what happens if a supplier does not generate/buy enough power for its customers? As the grid is fully interconnected, these customers will draw power from all other suppliers, causing instability for all customers. Clearly, some independent authority must be able to monitor production and demand for all operators, and ensure the proper balancing of production and demand.

Most countries have set up such an independent authority, the transmission system operator (TSO, see Figure 15.3 for a list of some European TSOs), which monitors in real time the production level of all operators connected to the grid, and the aggregate demand of all customers (see Figure 15.2, IESO forecasting demand in Ontario every five minutes).

As each operator is supposed to make sure its power production balances the demand of its customers through production planning and demand response, in an ideal world the actual aggregate production level should always match the actual aggregate demand. In reality, power consumption depends a lot on the weather (temperature, sun exposure), which is not completely predictable. In addition, some operators will sometimes have an issue that makes it very hard for them to meet the required production level: it might be an outage, a maintenance operation affecting some of their production facilities, or any unusual variation of actual demand compared to the projected demand. For all these reasons, actual demand will always differ slightly from the actual level of production.

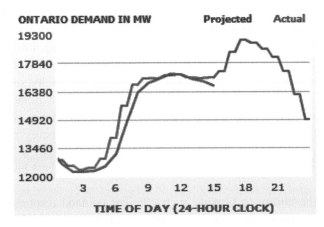

Figure 15.2   Demand projection by Ontario's IESO authority (independent electricity system operator), and actual demand.

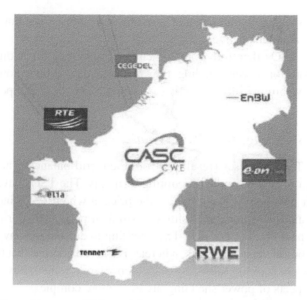

Figure 15.3 Some European TSOs and CWE, a cross-border capacity allocation agency.

The TSO in charge of balancing the network monitors in real time the production level of all power plants (and energy imports), as well as power consumption (using measurement systems along the main distribution lines). It also monitors the network frequency, which is the best instantaneous indicator of the production/demand imbalance: if the frequency is lower than the reference frequency (50 Hz in Europe), consumption exceeds supply, *vice versa* when the frequency is too high, supply exceeds demand.

The TSO has several tools to balance the network:

- Impose automatic production compensation algorithms to all major production plants: if the measured grid frequency is below target, the power plant will increase its power injection, and *vice versa*.
- If demand exceeds supply, the TSO can ask producers to increase their production (either as part of their contract, or by purchasing this energy), import energy from neighboring countries, or reduce demand (activation of power-shedding bids).
- If supply exceeds demand, the TSO can export energy, or sell excess energy to the best bidders. It can also ask producers to decrease their production.

The TSO expenses associated to this continuous balancing activity are funded by the electricity operators, usually *pro-rata* of the actual unbalance between their power production (own or acquired), and the power consumption of their customers.

The production capacity mobilized at the last minute by TSOs is often very expensive, and can even get prohibitively expensive in the case of major unplanned outages. In addition, this peak power production is also extremely inefficient in terms of CO_2 generation.

Again, asking users to shed power demand is an attractive alternative, both financially and from an environmental perspective.

More and more TSOs therefore complement their auctions for additional power by symmetrical auctions for power shedding: they receive bids to reduce power demand, and select all competitive offers instead of activating or sourcing additional power production.

15.4.2 Power Shedding: Who Pays What?

The allocation of TSO expenses (resulting from their grid-balancing activities) to each commercial operator is not a simple accounting formality. The TSOs have a lot of margin to implement their own policy. For instance, the price at which TSOs buy excess energy from operators in a positive imbalance situation (in their perimeter) during a grid power-deficit period is a pure matter of policy: TSOs willing to discourage any form of random imbalance should buy at very low prices, whereas TSOs considering that the operator contributed to the network stability (albeit by coincidence) may want to buy at a higher price.

The financial aspects of power shedding are even more complex:

- Companies specialized in demand management may be distinct from commercial suppliers, and are really a new type of player.
- The best policy from a financial fairness point of view may not be the optimal proposal from an environmental point of view
- The best power-reduction proposals may come from customers of other operators than the operators that caused the network unbalance. For each MW not consumed by these customers participating in the demand response scheme, their electricity supplier will not be charging its retail price R, and at the same time is still producing the electricity it had (correctly) planned its users would demand if the demand-response mechanism had not unexpectedly changed this demand profile. Therefore, the operator of customers participating in the demand-response program is still spending the then current marginal cost of power production M, but no longer charges R.

Obviously, the commercial suppliers of customers participating in the demand-response program will want to be indemnified for the additional production that contributed to the grid balance but was not charged to the customers.

TSOs and regulators need to decide whether there should be an indemnification, the level of indemnification, and who pays what. Obviously the arguments of each player will reflect their own interest:

- We will show below that the overall turnover of commercial suppliers is largely unaffected by demand response (the lower consumption is temporary and usually compensated by higher consumption later on). However, we will also see that these operators do produce more energy than what would be otherwise required, and their costs increase. Commercial suppliers will claim that their business is dependent on the accuracy of generic demand profiles, and since the additional production costs are caused by demand-response operators, these operators should compensate the additional costs.

- Some demand-response companies will claim that they have no relationship with the commercial suppliers; they are only dealing with the final customers who delegated their energy management to them, and therefore have no reason to compensate the additional costs of the operator.

In fact, these choices are more political choices than network management matters. The rapidity of the migration of energy networks towards "smart grids", and the resulting environmental impact will largely depend on the choices made.

We will study in the next section the principles and business case of demand response on a real network (the French grid managed by TSO RTE), in order to highlight the critical role of standards such as ETSI M2M in the future of such smart-grid applications.

15.4.3   Automated Demand Response

Demand response can apply to any process that can dynamically modify its consumption (or production) level. Many industries, for example, fresh water distribution (Figure 15.4), are ideal candidates for demand response because they can adjust their power consumption very quickly, and have important storage capacity.

Such industrial facilities typically use M2M technologies to retrieve real-time state information (in the example of Figure 15.4: reservoir level and pump power), which is used as an input for an optimization algorithm that manages demand-response participation. M2M is also used to transmit commands (e.g., starting or stopping a pump) during execution of the demand-response program.

In the following sections, we will study one particular way of participating in demand-response programs, by controlling the setpoint of heating/cooling systems in individual homes.

15.5   Case Study: RTE in France

15.5.1   The Public-Network Stabilization and Balancing Mechanisms in France

The French national consumption represents about 490 TWh (2008). The French power system has an installed capacity of over 118 GW, and typically provides 80 GW to over 90 GW at peak loads. Demand load depends a lot on the weather: in autumn, winter or spring, a temperature drop of 1 °C results in additional loads up to 2100 MW on the French network (data from RTE, 2009), while in summer, when the temperature exceeds 25 °C, a rise of 1 °C may bring about an extra load of up to 600 MW (data from RTE, 2004).

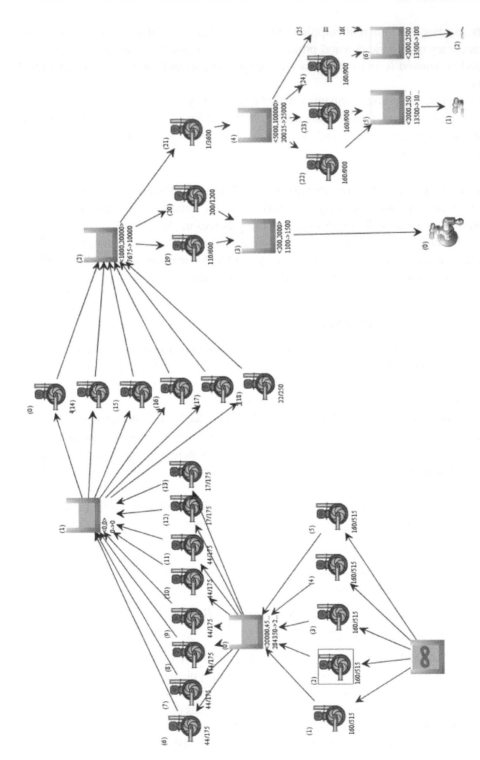

Figure 15.4 Model of a water-distribution system used for automated demand-response programs. Courtesy Actility, Smart-DR screenshot.

In France, the grid supply and demand adjustment became the responsibility of an independent authority, RTE (Réseau Transport Electricité), in 2002. Each French or foreign commercial electricity operator connected to the public grid must report their production, their projected and actual demand to RTE. They are also called "balance responsible entities", and are responsible for properly balancing supply and demand within their own perimeter. Each balance-responsible entity may use its own production facilities, or buy in advance power production from others. RTE monitors in real time the balance of the grid and the load of main transmission lines, and takes the appropriate short-term actions to make sure the aggregate supply and demand are balanced, and that the load on transmission lines remains within acceptable limits.

During exceptional network imbalance events, RTE may put in place special network-protection measures, such as a 5% drop of tension or in extreme cases disconnection of high-voltage transformers, network split-up, and so on. Of course, such last-resort options must be avoided and remain exceptional. In order to adjust supply to the level of demand, RTE uses three levels of routine regulation mechanisms, which are permanently in place and used during normal network operations: the primary, secondary and tertiary adjustment mechanisms (Figure 15.5).

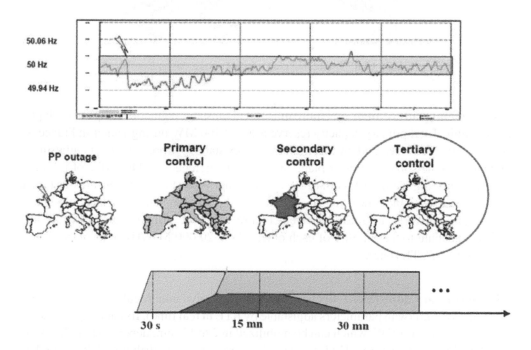

Illustration courtesy of RTE

Figure 15.5 Restoring balance: from immediate (primary) to midterm (tertiary) compensation mechanisms.

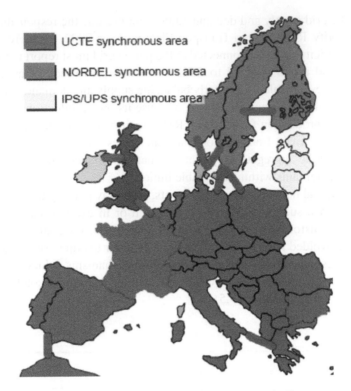

Figure 15.6   Interconnected synchronous areas in Europe.

The **primary adjustment mechanism** is based on static regulation loops in each production unit. Total primary capacity reserve is about 700 MW during winter in France. If, as a result of excess demand, the network frequency starts to drop, the primary adjustment mechanism reacts in each power plant, and adds an aggregate of 4400 MW of power or more per missing Hertz in less than 30 s. The primary power reserve ensures fast local regulation close to the power injection point. As this mechanism is in place throughout Europe and networks are interconnected (Figure 15.6), the primary reserves of all networks contribute to the stability of each other (about 20 000 MW/Hz of primary reserve power in the UCTE zone).

- The **secondary adjustment mechanism** is based on dynamic regulation loops in each production unit (with parameters adjustable by RTE in real time). Its capacity represents about 500 to 1000 MW, which can be mobilized in 2 to 15 min, depending on the extent of the adjustment required. Hydro power plants are a major contributor of the secondary reserves. Secondary-level adjustments provide coordinated regulation over wider areas, and aim at maintaining frequency stability and network synchronicity, while restoring the primary reserves.

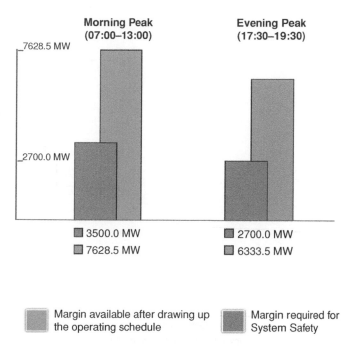

Figure 15.7 RTE uses the tertiary adjustment bids to maintain acceptable supply margins during demand peaks.

- The **tertiary adjustment mechanism**, based on semiautomatic and manual procedures, has a capacity of about 1000 MW within 15 min and an additional 500 MW within 30 min. The French demand-response market supplies the tertiary adjustment mechanism, and is used to readjust the daily power-generation schedule in order to reconstitute the primary and secondary adjustment capacities (Figure 15.7).

15.5.2   The Bidding Mechanisms of the Tertiary Adjustment Reserve

The management of the tertiary adjustment reserve is based on a permanent bidding mechanism in order to maintain an up to date inventory of potential short-term balancing offers: additional production or power shedding offers in case demand exceeds supply (see Figure 15.8), supply reduction or demand increases in case supply exceeds demand.

Bids for day D are open at D–1 4 p.m. and 9 p.m., and then can be added, modified or removed on D day every hour.

Each bid is characterized by:

- Its nature. Additional supply bids and power shedding bids allow RTE to compensate for an excess of demand: RTE will pay for the additional power injected in the public

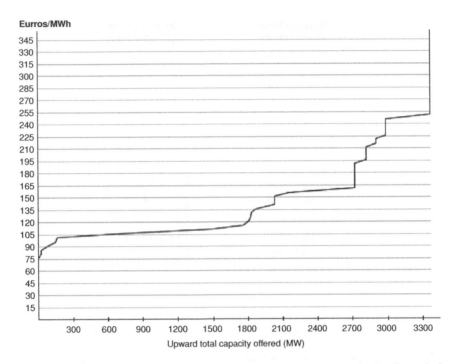

Figure 15.8 RTE typical cumulative upward bid capacity and marginal prices during the 17:30–19:30 peak time.

grid, and for the demand reduction of bidders. Supply-reduction bids and demand-increase bids allow RTE to compensate an excess of supply: the bidders will pay for the electricity sourced from the public grid, which they use either to replace their own supply reduction, or to power the extra demand.

- A validity period.
- A selling price (which might be a time-dependent price).
- An activation delay: the time that the bidder needs to make its offer effective and compliant with RTE activation level after it receives an activation notification.
- A minimal activation time.
- The amount of corrective power available (capacity available in MWh and peak power in MW), with a minimum of 10 MW for a half-hour. The corrective power may consist of additional supply, power shedding, supply reduction or demand increase depending on the nature of the bid.

In addition, for production plants participating in the adjustment power bidding, RTE requires to receive on D−1 the normal production/demand plan for day D, in order to check the effectiveness of additional supply bids.

The public network balancing state is assessed in real time by RTE. The network is self-stabilized by the primary adjustment and secondary mechanisms described above

(Figure 15.5), but RTE needs to restore the margins of these immediate reserves by using the corrective capacity made available by the bidders:

- **If demand exceeds supply:** RTE selects additional power-supply and power-shedding bids (what RTE calls the "fast reserve" of power) in increasing price order, and activates them totally or partially. The level of activation can be adjusted in real time by RTE within the limits of the proposed bid.
- **If supply exceeds demand:** RTE selects supply-reduction and additional demand-purchasing bids in decreasing price order, and activates them totally or partially.

15.5.3   Who Pays for the Network-Balancing Costs?

As illustrated in Figure 15.9, in a typical day RTE will activate more power to compensate for excess demand (paying prices above the spot price) than to compensate for excess supply (charging the sold excess power below spot market prices). As a result, this stabilization mechanism represents a net cost.

RTE decided to allocate this cost among equilibrium operators who contributed to the network imbalance, as illustrated in Figure 15.10.

RTE administratively classifies every half-hour in the day in the "upward adjustment" (excess demand) or "downward adjustment" (excess supply) category, based on the dominant trend in the half-hour.

If demand exceeded supply during the past half-hour (Figure 15.10, top):

- Some "balance responsible entities" may be in a situation where supply in their perimeter exceeded demand for this half-hour, effectively reducing the aggregate network

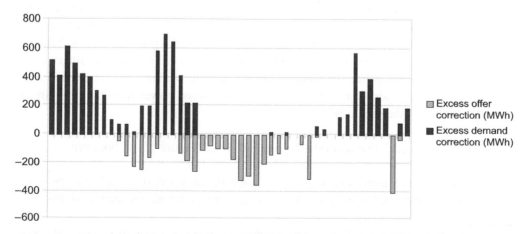

Figure 15.9   Typical RTE daily activation profile of adjustment bids (in half-hour steps).

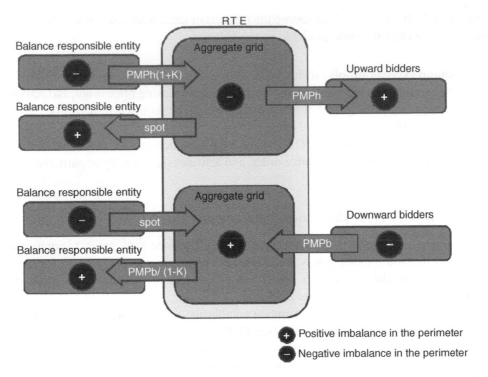

Figure 15.10 Financial flows between RTE, balance responsible entities and bidders, depending on the balancing state of their respective perimeters.

imbalance: in this case RTE buys the excess supply at the Powernext EPEX spot price determined at D–1 for delivery in this half-hour.

- The total balancing cost for the half-hour PMPh[1] (activated bids and purchases at the spot price), adjusted by a deterrent coefficient K (about 1.05), is shared among the "balance responsible entities" who were net users of the public network power during the half-hour, *pro-rata* of their actual use of the public grid resources to balance their perimeters. Since using these resources is costly, this of course encourages each "balance responsible entity" to properly predict demand within its perimeter in order to buy additional production in advance, rather than draw power from the last-minute "fast reserves" of the public grid.

If supply exceeded demand during the past half-hour (Figure 15.10, bottom):

- Some "balance responsible entities" may be in a situation where demand in their perimeter exceeded supply, effectively reducing the aggregate network imbalance: in this case

---

[1] For "Prix Moyen Pondéré à la Hausse": average weighted price of increase bid offers (additional power, demand reduction) that were activated.

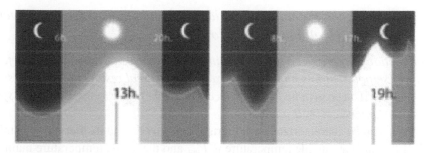

Figure 15.11   Typical demand profile during summer and winter: peak hours.

RTE provides the required additional supply, charging the Powernext EPEX spot price determined at D–1 for delivery in this half-hour.

- The total balancing revenues for the half-hour PMPb[2] (activated bids and power sales at the spot price), adjusted by a deterrent coefficient $1/K$, is shared among the "balance responsible entities" who were net suppliers of the public network during the half-hour, *pro-rata* of their actual supply. Since the purchasing price is below the spot price, this encourages each "balance responsible entity" to properly predict demand within its perimeter in order to sell excess production in advance, rather than below the spot price at the last minute.

15.6   The Opportunity of Smart Distributed Energy Management

In the past, only industrial sites were bidding as part of the demand-shedding mechanism. In addition, the residential and SME demand factor have been modulated by flexible tariffs[3] in order to reduce demand during predicted peak hours, and in the most sophisticated cases, during predicted peak days (Figure 15.11). But these tariffs gradually disappeared after the liberalization of the electricity markets, due to the fact that not all operators could send the necessary tariff change triggers over the electricity network. In France for instance, it has been estimated that about 150 MWh of demand-shedding power disappears every year.

But the fact that most homes are now connected to the Internet, and the relatively low cost of the systems required to regulate energy usage in the home, are changing the situation. Since 2008, RTE experiments distributed power-shedding mechanisms using the Internet as the control network. Such a system could potentially be extended to virtually all homes and businesses using electrical heating, as broadband Internet penetration is very high in France.

---

[2] For "Prix Moyen Pondéré à la Baisse": average weighted price of decrease bid offers (power reduction, demand increase) that were activated.

[3] Tarif EJP ("Effacement des jours de pointe" or peak day demand-shedding tariff. It is no longer sold to new customers.

15.6.1 Assessing the Potential of Residential and Small-Business Power Shedding (Heating/Cooling Control)

For simplification, we consider a perfect heating system that exactly maintains its reference temperature (setpoint) at any moment. In practice, most real residential heating systems are either switched on or switched off, with a certain temperature hysteresis. These systems are studied in Appendices A, B and C.

Each house can be characterized by it thermal capacity C, and its thermal transmission factor K. If the current outside temperature is T_{out} and the inside temperature maintained by the heating system is T_{ref}, then the average heating power consumed by the house is:

$$P = K \times (T_{ref} - T_{ext})$$

The demand shedding mechanism will reduce T_{ref} by a small amount Δ, the new reference temperature will become $T_{ref} - \Delta$. The heating system will switch off to let the house cool to the new desired temperature. This will last $(C \times \Delta)/[K(T_{ref} - T_{ext})]$ (first-order approximation if $\Delta << (T_{ref} - T_{ext})$, the exact formula for the temperature evolution is an exponential).

After this cool-off period, the heating system is turned back on, and the power consumption becomes:

$$P' = K(T_{ref} - \Delta - T_{ext})$$

For a typical house, let us evaluate the order of magnitude of these parameters during winter ($T_{ref} - T_{ext} = 10 \,°C$), for a power-shedding period of 3 h allowing a decrease of T_{ref} by 1 °C ($\Delta = 1 \,°C$):

- Average heating power $P = K^*(T_{ref} - T_{ext}) = 3$ kW
- Time to cool by 1 °C when heating is switched off :

$$C \times \Delta/K(T_{ref} - T_{ext}) = 100 \text{ min}$$

- Energy saved during the cooling period = 3 kW × 100 min = 5 kWh. The same amount of additional energy will need to be consumed later in order to reheat the house back to T_{ref}.
- Energy saved during the remaining 80 min (1.33 h) at a lower temperature of $T_{ref} - \Delta$:

$$K \times \Delta \times 1.33 = P \times \Delta/(T_{ref} - T_{ext}) \times 1.33 = 0.4 \text{ kWh}$$

This saving, which reflects the lower level of comfort accepted by the home owner during the power-shedding period, will never be compensated by higher consumption later on if the heating setpoint is reset to its original value T_{ref}.

For 5000 such houses, the energy saved during the power shedding period is about 27 MWh. In order to reheat the house back to T_{ref}, about 25 MWh of higher consumption will have to be planned later on.

This simplified analysis does not take into consideration the hysteresis of the heating systems. In practice, the temperature-evolution curve in most homes is a zig-zag function of time (Appendix A, Figure A.1): when the heating system is on, the temperature increases by X °C/h, when it is off, it decreases by Y °C/h. A traditional thermostat regulates the temperature with a hysteresis of H °C: the heating system is switched on when the temperature reaches T_{ref}, and switched off when it reaches $T_{ref} + H$. We take this hysteresis into account in Appendix A and show in Appendix B that, as a result of this hysteresis, it is possible for a demand-management system to introduce synchronization across the heating systems, that is, they will all switch off and on at the same time, instead of randomly. This can create unacceptable consumption peaks in the grid. Therefore, the exact thermostat control algorithm is important: Appendix C exposes a control law which does not introduce synchronization, but instead introduces gradual power decreases and increases at the beginning and end of the power-shedding period. These issues are the main motivation for the introduction of randomization factors in ZigBee SE 2.0 (see chapter 13).

15.6.2   Analysis of a Typical Home

Figure 15.12 shows the values for X, Y and H for a typical home, and Figure 15.13 shows the daily power consumption as a function of external temperature. When active, the

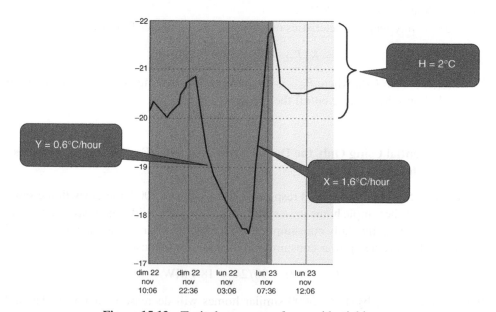

Figure 15.12   Typical parameters for a residential home.

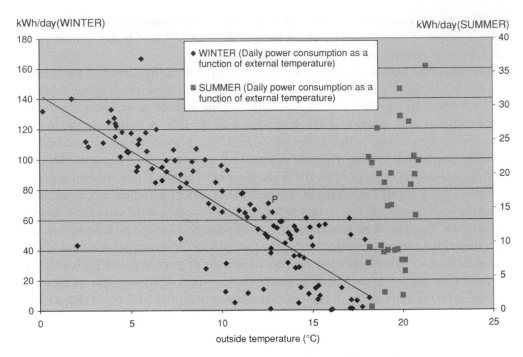

Figure 15.13   Daily power consumption as a function of external temperature.

electric heating system is clearly dominant over other uses. The yearly power usage of this 120 m^2 house is about 20 MWh.

The various physical characteristics of the home are related as follows:

$$Y = K(T_{\text{ref}} - T_{\text{ext}})/C \approx 0,6°C/h$$

$X = P_{\text{max}}/C \approx 1,6°C/h$, where P_{max} is the power of the heating system when turned on (this sample house has a basic on/off thermostat).

15.6.2.1   Potential Using Only the Daily Excess Demand / Excess Supply Volumes on the Public Grid

Let's evaluate the potential demand response contribution of 5000 homes (with the same characteristics as the sample home we studied above) during a mild winter day: the outside temperature is 10 °C, the daily consumption is 90 kWh in order to maintain about 20 °C in the home. The average power consumption of these 5000 houses is:

$$P = 5000 \times 90/24 = 18.7\,\text{MW}$$

If we decrease T_{ref} by 1 °C, 5000 similar homes will decrease their power demand from 18.7 MW to zero in 1 h 15 min (H/X, see Figure 15.14 and Appendix C where we

Δ	Tdown (100% to 0% P in h)	Tzero (0% P in h)	Tup (0% to 100% P in h)	Tzero+(Tdown+Tup)/2 Equivalent time at 0% P
−1°C	1:15	00:25	1:15	1:40
−2°C	1:15	2:05	1:15	3:20
−3°C	1:15	3:45	1:15	5:00

Figure 15.14 Demand response cycle parameters with $H = 2$, $X = 1.6$ and $Y = 0.6$.

use a demand-response policy that avoids synchronization effects), then power demand will remain null for an additional 25 min (Δ/Y–H/X), and increase back to 18.7 MW in another 1 h 15 min (in reality the power consumption will be slightly less, due to the lower thermal dissipation when the house is 1 °C colder, see Appendix D for details). The full cycle saves 31 MWh compared to the normal consumption pattern and takes about 3 h to complete, leaving all homes 1 °C colder.

5000 120 m² homes, winter day (10 °C outside temperature)
−1 °C → −31 MWh over 3 h

In order to maintain the average home temperature settings, such a −1 °C negative demand response cycle would need to be followed by a reheat cycle of +1 °C: this cycle would raise power demand from 18.7 to 28 MW ($P \times (H+\Delta)/H$, see Figure 15.15 and annex C) in 37 min (Δ/X), then remain at 28 MW for 2 hours 42 min (H/Y–Δ/X), then decrease to 18.7 MW in another 37 min. The full reheat cycle drains an additional 31 MWh compared to the normal consumption pattern, and takes about 4 h to complete.

5000 120 m² homes, winter day (10 °C outside temperature)
+1 °C → +31 MWh over 4 h

We can see that, **if the energy tariff is flat, the impact of a full power-shedding / reheat cycle on the home owner energy bill is nearly neutral:** if we neglect the slightly lower thermal dissipation of the house during the power-shedding period the energy saved during the power-shedding period is the same as the additional power consumed during the reheat period. However, the electricity tariff is rarely flat for homes using electric heating,

Δ	Tup (0% P to P*(H+Δ)H in h)	Thigh (Demand=P*(H+Δ)H in h)	Tdown (P*(H+Δ)H to 0% P in h)
+1°C	0:37	02:42 (150%P)	0:37
+2°C	1:15	2:05 (200%P)	1:15
+3°C	1:52	1:27 (250%P)	1:52

Figure 15.15 Reheat cycle parameters with $H = 2$, $X = 1.6$ and $Y = 0.6$.

therefore electricity adjustment frameworks should seek to decorrelate the metering data used for billing purposes by the commercial supplier and the metering data used for the demand response participation. This is the case is France for industrial demand-response (>250 kW) where RTE automatically compensates metering data before sending it to commercial suppliers. The exact mechanism is still under discussion for lower power demand response use cases.

In order to evaluate the maximum potential of demand response in a favorable case, we are taking the following maximum tolerance assumptions of home owner for changes of the home minimal temperature (i.e. Δ), *in addition to that already planned* (Figure 15.12 shows that a -3 °C temperature drop is already planned at night for our sample house), depending on the time of day:

- between 7 a.m. and 9 a.m.: -1 °C (family wakes up, breakfast time);
- between 9 a.m. and 11:30 a.m.: -2 °C (children at school);
- between 11:30 a.m. and 2 p.m.: -1 °C (lunch time);
- between 2 p.m. and 6:30 p.m.: -2 °C (children at school);
- between 6:30 p.m. and 10 p.m.: -1 °C (dinner, going to bed);
- between 10 p.m. and 7 a.m.: -2 °C (everybody sleeping).

The home-owner allowance is at least -1 °C for the entire day: therefore over 24 h there could be up to 3 such negative demand-response/reheat cycles (7 hours for each cycle), totalling 93 MWh of negative demand response for 5000 similar homes (and 93 MWh of additional demand outside of the negative demand-response slots). It is rather difficult to use the additional allowance of -2 °C, because at a minimum 7 h are necessary to decrease the temperature by 1 °C and get back to normal. This seems possible only at night between 10 p.m. and 3 a.m. (the 3 a.m. to 7 a.m. period is required for the reheat cycle). This would add another 31 MWh of negative demand response between 3 a.m. and 7 a.m. (and 31 MWh of additional demand outside of the negative demand-response slots), bringing the total demand response potential to about 124 MWh for 5000 homes with the same characteristics as our sample home, assuming a "very" tolerant home owner.

Of course, ideally such demand-response/reheat mechanisms should always contribute to a better network equilibrium. This is possible only if excess demand and excess supply periods alternate during the day: fortunately, as illustrated in Figure 15.9, this is the case. However, as illustrated in Figure 15.16, the volume of upwards demand adjustments is often less than the volume of downwards adjustments.[4] In the current RTE mechanism, the realistic maximum volume of demand response would need to be lower than about 1000 MWh per day in order to be compensated within 24 h during the daily excess supply periods. With 2 demand-response cycles of -1 °C and one cycle of -2 °C per day, this represents only about 40 300 homes!

---

[4] See http://clients.rte-france.com/lang/an/clients_traders_fournisseurs/vie/mecanisme/histo/volume_type_offre.jsp.

MWh/day

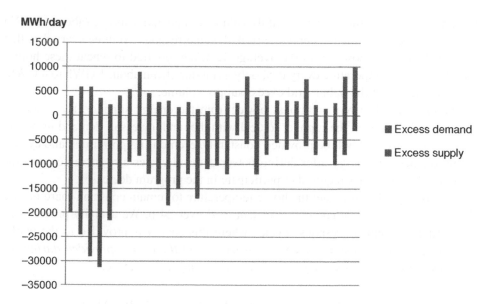

Figure 15.16 Daily volumes of excess demand and excess supply (September 2009).

If we had the possibility to get back to the normal temperature over a month, not within a day, then we would be theoretically limited only by the average monthly excess supply (about 4000 MWh/Day), not by the minimum daily excess supply. This makes the addressable demand response volume 4 times higher, but requires about 4×30 more homes (We need 4 times more energy, and 30 times more homes because it is possible to use only a 1/30 fraction of the daily temperature drift allowance per home every day in order to readjust over 30 days).

---

5 000 000 120 m² homes, winter day (10 °C outside temperature)

-1 °C demand response cycles, FULLY compensated during public grid excess supply periods $\leftrightarrow$ 4000 MWh of demand response per day (during winter).

15.6.2.2   Potential of Demand-Response Schemes that Modify the Average Demand

The demand-response potential that we evaluated above was limited by the volume of excess supply as part of the daily or monthly network-balancing bids, because we did not want to change the supplier production profile. However, if we have enough homes participating in the system and aim to fully address the monthly demand-response volume, we need to also modify the average power demand and therefore also the supplier-production profile.

We can address fully the average monthly volume of demand response (about 12 GWh per day) only if we are able to supply the peak demand-response volume every day, that is about 30 GWh. But now, out of the average 12 GWh required to reheat these homes back to the normal temperature every day, only a fraction (from about 1 GWh to 4 GWh) can be recovered from the public grid excess supply volume.

We therefore need to buy additional power from the energy supplier, for a total of αP. Since we do not want the supplier to deal with the randomness of demand response, this increased power αP will be provided constantly, even during power demand cycles: as a consequence our energy offer during power-demand cycles also increases, and the minimal number of homes required to participate in the program decreases.

Over an average day, we want the home temperature to remain constant, therefore we must provide an amount of energy equivalent to P over 24 h. We actually provide αP, except during the demand response cycles where this energy is provided to the public grid, not to the homes. We also recover a *fraction s of the energy provided during the demand response cycles from the public grid excess supply*. Therefore:

$$\alpha P \ (24 + (s - 1) \ [\text{total duration of demand response cycles}]) = 24 \times P.$$

$$\alpha = 24/(24 + (s - 1) \ [\text{total duration of demand response cycles}]).$$

Given N_1 the number of $-1\,^\circ$C power-shedding cycles per day, and N_2 the number $-2\,^\circ$C power-shedding cycles per day, Figure 15.17 evaluates the necessary increase factor

Scenario	Number of homes required to fully balance the network (12000 MWh/day with peaks to 30000 MWh/day)			Required additional base power (α–1) with the minimal number of homes required		
	s= 0%	s=10%	s=33%	s=0%	s=10%	s=33%
N1=1, N2=0	4 470 000	4 500 000	4 580 000	+7,5% P	+6,7% P	+4,9% P
N1=2, N2=0	2 070 000	2 100 000	2 180 000	+16% P	+14% P	+10% P
N1=3, N2=0	1 270 000	1 300 000	1 380 000	+26% P	+23% P	+16% P
N1=2, N2=1	8 70 000	9 00 000	9 80 000	+38% P	+33% P	+22% P

Figure 15.17 Minimal number of participant homes and additional base power required as a function of the demand-response scenario.

of the base power production, as well as the number of homes required to participate in the demand response program in order to fully balance the network.

15.6.3   The Business Case

15.6.3.1   The Residential and Tertiary Sector in Numbers

Overall, the energy consumptions of the residential and tertiary sectors represent about 47% of the total energy consumption in France,[5] compared with 28% for the industry and agriculture sector and 25% for the transports sector. The residential and tertiary sectors account for about 123 million tons of CO_2 emissions each year, or about a quarter of the total French CO_2 emissions. The residential sector alone accounts for about 2/3 of the total, while the tertiary sector accounts for only 1/3.

The business sector represents 850 million square meters (about 10 million equivalent homes, half public and half private), which are heated or cooled. The average energy efficiency varies widely depending on the sector, from about 131 kWh/m^2 in education to 322 kWh/m^2 in the transport sector.

The residential sector in France represents about 26 million homes totaling 3.5 billion square meters:

- about 13 million individual houses (including 2 million second homes and vacant homes);
- about 6 million condominiums (including 2 million second homes and vacant homes);
- about 4 million homes, apartments or houses, owned by institutional investors.

Most homes are still far from being energy efficient (see Figure 15.18): although there has been much progress, the average primary[6] energy usage which was about 372 kWh/m^2 year in 1973 is still about 240 kWh/m^2 year for heating (hot water included). This represents an average CO_2 emission of 35 kg/m^2/year. Despite the improvements in energy efficiency, the final energy demand in homes increased by 24% between 1973 and 2004. The 2008 statistics of the mandatory energy performance diagnostic ("Diagnostic de Performance Energétique"[7]) show the following distribution:

- 31% of homes (most homes built between 1975 and 2000) consume between 150 and 230 kWh/m^2 year of primary energy (energy efficiency class D).

---

[5] See http://www.ifen.fr/acces-thematique/activites-et-environnement/construction-et-batiments/construction-et-batiments/la-consommation-energetique-des-batiments-et-de-la-construction.html.

[6] For electricity, statistics consider that the primary energy represents a multiple of the final consumption in kWh. France applies a multiple of 2.58 (2.58 kWh of primary energy is required to deliver 1 kWh to the home). Other countries of energy-efficiency labels use other values, 2 for Minergie (Switzerland), 2.85 for Passivhaus (Germany). See also: http://www.fiabitat.com/labels-basse-energie.php.

[7] For a DPE simulator, see http://www.outilssolaires.com/Archi/prin-perf.htm.

kW/m²/year	% of total	Average energy bill per room (Ile de France, 2008)
<= 50 (Class A)	–	<35€
51–90 (Class B)	–	35–63€
91–150 (Class C)	8%	63–106€
151–230 (Class D)	25%	106–162€
231–330 (Class E)	34%	162–232€
331–450 (Class F)	20%	232–316€
>450 (Class G)	12%	>316€

Figure 15.18 Energy efficiency of homes with electrical heating (air conditioning + sanitary hot water).

- 22% of homes consume between 230 and 330 kWh/m² year (class E).
- 18% of homes (most homes built after 2000) consume between 90 and 150 kWh/m² year (class C).

The sample house we studied above and used as a basis for our evaluations has a yearly consumption of 20 000 kWh, and 20 kWh/day for specific electricity uses, including about 10 kWh for sanitary hot water. It therefore uses about 16 000 kWh of electricity for heating and sanitary hot water, which represents about 350 kWh/m² of primary energy (Class F).

The typical uses of energy at home are listed in Figure 14.19. The relative weight of heating is expected to decrease significantly over the next 40 years, with the ever-increasing number of powered appliances in our homes, and the slow penetration of better energy-efficiency standards. The power consumption linked to IT and telecoms alone increases by about 10% per year, representing about 15% of the total consumer bills.[8] A single Internet access device consumes over 50 kWh per year, and sometimes much more. Although the unitary consumption of these appliances improves generation after generation, there are more and more such devices in every home.

Like our sample home, 30% **of principal residences in France (about 8 million) use electrical heating**. In France, during the winter electric heating is competitive on average: it releases about 180 g of CO_2 per kWh of heat due to the use of nuclear energy, compared to 300 g for fuel heating and 234 g for gas heating (source Ademe). But during the peak periods, traditional power plants are used and electric heating generates 500 to 600 g of

---

[8] OCDE estimate (Christian REIMSBACH-KOUNATZE) is 15%, government estimate (rapport DETIC http://www.telecom.gouv.fr/actualites/11-mars-2009-rapport-sur-les-tic-developpement-durable-2045.html) is about 13.5% (55–60 TWh). The same study estimates that IT systems help save about 4 times the amount of CO_2 they generate.

Home energy usage	Today (240 kWh/m^2.year home)	2050 (60 kWh/m^2.year)
Heating	87%	30%
Hot water About 855 kWh/person.year, or about 25 kWh/m^2.	6%	30%
Cooking	3%	10%
Other	4%	30%

Figure 15.19   Energy uses in the home, now and in 2050.

CO_2 per kWh of heat. It is therefore very important to try to reduce the peak times as much as possible!

15.6.3.2   Demand Response: The Network-Balancing Business Case

5000 Homes, Single Power-Shedding Cycle per day (−1 °C), Reheat During Excess Supply

In the previous sections, we showed that by properly regulating the thermostats of about 5000 homes, with a single decrease of T_{ref} per day followed by a reheat cycle, we could bid for about 31 MWh/day as part of the network-balancing power-shedding bids, and reheat the house during the excess power supply periods (this ensures that the energy producer does not need to deviate from normal energy-production planning).

We will evaluate the revenues derived from the power-shedding bids using an average price per MWh for the activated power shedding bids of 80 €/MWh (a bit lower than the actual 1st of November to March 31st average in 2008).

The cost side is more subtle. Let us consider the normal situation (Figure 15.20), and the situation during a demand-response cycle (Figure 15.21), from the point of view of

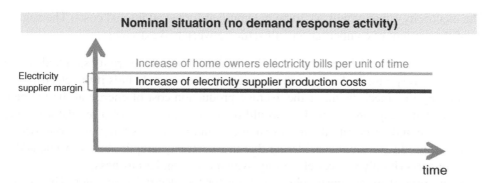

Figure 15.20   Home owners bills and supplier costs without demand response.

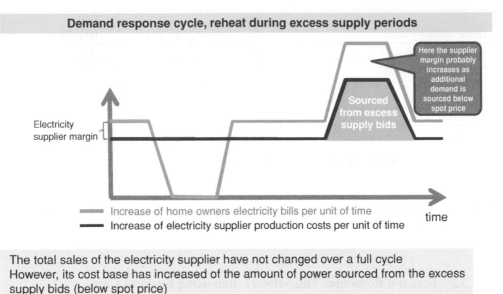

The total sales of the electricity supplier have not changed over a full cycle
However, its cost base has increased of the amount of power sourced from the excess
supply bids (below spot price)

Figure 15.21   Home owners bills and supplier costs during a demand-response cycle.

financial flows. In our simplified model the production cost of electricity is constant over
time.

The total revenues of the energy supplier remain identical to the revenues it would have
had without demand response (we neglect the effects of the variation of the home temperature on its thermal dissipation, these effects are evaluated in Appendix D). However,
while the entity who organized the demand-response bids earns the bids revenues, the
electricity supplier sees its costs increase, because it needed to source the reheat energy
(or even worse . . . reheating may have caused a negative imbalance in its perimeter). Here,
we further suppose that the company managing the power-shedding scheme is exchanging
information with the energy supplier so that this energy can be purchased through excess
supply bids.

The electricity supplier will ask to be indemnified for the additional costs. The negotiation of the amount of compensation will settle between two extremes:

- Exact compensation of the excess costs, that is, the volume of electricity supplied during
  the demand-response cycle, at the mean price of excess supply bids if the electricity is
  supplied by this mechanism, or the average production cost of energy at the moment
  of the demand response event. This would be the point of view of a regulator willing
  to maximize and accelerate the use of such demand-response strategies as opposed to
  additional production, and considering that the electricity supplier has no value added
  in this process (it only sources electricity, which is its regular business).
- Compensation of excess costs, plus a margin for the electricity supplier. This would
  be the default situation without regulatory pressure. In an extreme case the electricity

supplier could be willing to charge its regular residential tariffs, possibly higher than most demand response bids, severely impacting the business case for residential demand response.

Who should pay? From a purely *economic point of view*, it seems logical that the company who caused the additional cost (the power-shedding company) should pay for it. From a *legal point of view*, this is not totally clear as power-shedding companies may claim that they are only mandated by their customers to better regulate energy use in their homes, and customers never signed a contract forcing them to indemnify the operators for nonstandard demand profiles. From an *environmental perspective*, it might be optimal to consider that this additional energy supply is actually part of the grid-balancing system "costs", due to a preference for demand shedding over additional production caused by the environmental constraints: in this case, it would ultimately be reflected by an increase of the K coefficient (see Section 15.5.3), and paid by the operators.

In the business cases exposed below, we consider that the power-shedding company will be responsible for the indemnification, which is obviously a worst case from the point of view of the power-shedding companies. For now (September 2011), the compensation level has not been settled in France, which is why we evaluate several scenarios in the business cases.

If we consider a cold season of 5 months, the yearly bid revenue for these 5000 homes is about 377 k€ ($31 \times 80 \times 365 \times 5/12$). The net revenue per home is estimated in Figure 15.22.

In addition, if the commercial price of electricity is about 100 €/MWh, and considering the average heating power of our sample house (3.74 kW), the home owner also sees his

	Gross revenue per home (€/year)	Compensation of electricity supplier (€/year)	Revenue sharing with user (50%) (€/year)	Net revenue per home per year (€/year)
Compensation at mean price of excess supply bids ($\sim$40€/MWh)	75	−37	−19	19
Compensation at mean cost of base production ($\sim$50€/MWh)	75	−47	−14	14
Compensation at mean cost of base production, plus 20% margin ($\sim$60€/MWh)	75	−56	−9	9

Figure 15.22 Business case estimation, 5000 homes, single power-shedding cycle per day.

energy bill decrease by about 15 € (See Appendix D, we use the approximation $H = 0$ to evaluate the period during which the heating system is off to 1 h 40 min (Δ/Y), and we arbitrarily consider that the average period during which the heating system maintains $T_{\text{ref}}-\Delta$ is 2 h as we do not know exactly when the next reheat cycle will be).

40 300 Homes, 4 Power-Shedding Cycles (–1 °C) per day, Reheat During Excess Supply

This corresponds to the maximum potential of demand response with very tolerant home owners. We showed that by properly regulating the thermostats of about 40 300 homes, about 2 power shedding cycles of $-1\,°C$ and one power shedding cycle of $-2\,°C$ could be arranged, representing the maximum power-shedding capability of the home. With this strategy we can bid for about 1000 MWh/day as part of the network-balancing-power shedding bids, and still reheat the house during the excess power-supply periods.

The cold-season bid revenue for these 40 300 homes is about 12 M € (Figure 15.23).

In addition, if the commercial price of electricity is about 100 €/MWh, and considering the average heating power of our sample house (3.74 kW), the home owner also sees his energy bill decrease by about 60 €.

	Gross revenue per home (€/year)	Compensation of electricity supplier	Revenue sharing with user (50%)	Net revenue per home per year
Compensation at mean price of excess supply bids (~40€/MWh)	301	−151	−75	75
Compensation at mean cost of base production (~50€ /MWh)	301	−189	−57	57
Compensation at mean cost of base production, plus 20% margin (~60€/MWh)	301	−226	−38	38

Figure 15.23   Business case estimation, 40 300 homes, 4 power-shedding cycles per day.

Addressing the Full Power-Shedding Bidding Capacity

There are about 8 million homes in France using electric heating. We suppose that about a quarter of these homes can engage in a demand-response program: with these 2 million homes, and using about 2 demand-response cycles per day, we can fully balance the network and provide the required 12 000 MWh of demand response per day.

Scenario	Number of homes required to fully balance the network (12000 MWh/day with peaks to 30000 MWh/day)			Required additional base power (α−1) with the minimal number of homes required		
	s= 0%	s=10%	s=33%	s=0%	s=10%	s=33%
N1=1, N2=0	4 470 000	4 500 000	4 580 000	+7.5% Pn	+6.7% Pn	+4.9% Pn
N1=2, N2=0	2 070 000	2 100 000	2 180 000	+16% Pn	+14% Pn	+10% Pn
N1=3, N2=0	1 270 000	1 300 000	1 380 000	+26% Pn	+23% Pn	+16% Pn
N1=2, N2=1	870 000	900 000	980 000	+38% Pn	+33% Pn	+22% Pn

Figure 15.24 Selected demand response scenario for our business case.

However, we also need to increase the normal demand of these homes by about 14%. Also, in order to accommodate the daily peaks of demand response (30 000 MWh), we will use, on average, only 40% of the available demand-response capacity of each home (Figure 15.24).

Now, the financial flows are also different (Figure 15.25).

We have spread the reheating energy demand evenly so that the electricity producer could cover it with its base production (the production that can be planned well in advance, and therefore sourced at the best prices). Clearly, this level of production cannot be stopped during the random demand-response periods so, for the energy supplier production planning, everything looks as if our homes had increased their energy demand by about 14%.

However, this additional cost is not covered by the home owners: their total bill over a demand response/reheat cycle is still identical to what it would have been without demand response.

Again, the energy supplier must be compensated for the additional cost. This time, the minimal compensation cost will be the base production cost of the supplier. We can source a fraction of this electricity from the public grid during the excess supply periods, but overall this will not represent more than 10 to 20% of the total, so the impact on the average electricity price is negligible for our evaluation. Figure 15.26 shows the resulting business case, depending on the purchasing price of electricity supplied during demand response bids.

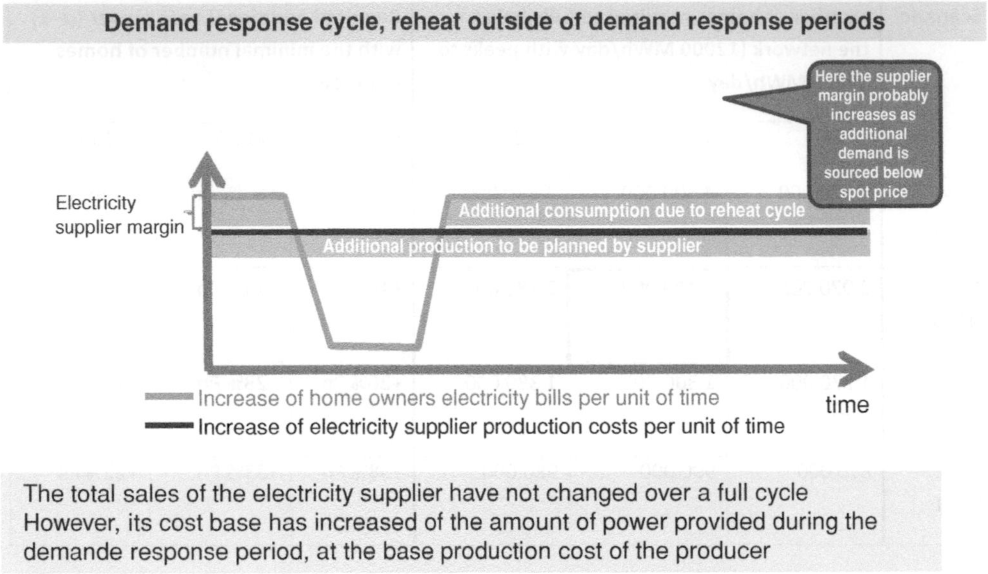

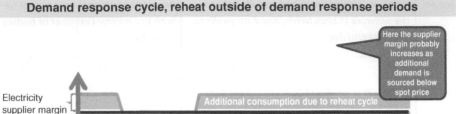

Figure 15.25 Home owners electricity bill and electricity supplier costs over a demand-response cycle (constant reheat outside of the demand-response cycle).

	Gross revenue per home (€/year)	Compensation of electricity supplier	Revenue sharing with user (50%)	Net revenue per home per year
Compensation at mean cost of base production (~50€ /MWh)	70	−43	−13	13
Compensation at mean cost of base production, plus 20% margin (~60€/MWh)	70	−52	−9	9

Figure 15.26 Business case estimation, 2 100 000 homes, 2 power-shedding cycles per day (12 000 MWh/day on average).

15.7 Demand Response: The Big Picture

15.7.1 From Network Balancing to Peak-Demand Suppression

15.7.1.1 Feasibility of Peak-Demand Suppression

On a typical winter day, demand will fluctuate by about 50% from peak to trough (Figure 15.27). The peak usually occurs at 7 p.m. and can cause serious problems, especially

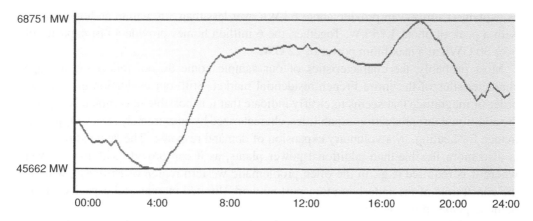

Figure 15.27   Typical daily demand profile (from http://clients.rte-france.com/lang/fr/clients_consommateurs/vie/courbes.jsp).

during the coldest days: -1 °C adds about 2100 MW to the national consumption, or twice the equivalent of the consumption of a large city for example, Marseilles (this sensitivity increases over time it was 1800 MW/°C in 2006 and 1500 MW/°C in 2000). Peak power consumption is in constant increase, and the trend is projected to continue : Figure 15.28 shows the forecast of RTE.

EDF, the incumbent national operator, decided to add, from 2006 to 2012, about 4000 MW of ultrafast thermal production capability (e.g., fuel turbines) that can start and provide power within 20 min. By comparison, an EPR nuclear reactor has a capacity of about 1600 MW. These turbines will be used typically less than a hundred hours per year.

The previous evaluations show that about 25% of homes using electric heating in France would need to participate in a demand-response program to completely balance the network during the winter.

It is interesting to evaluate also what could be achieved if participation in a demand response program became mandatory by law for all users of electric heating (this policy would have some logic, considering the need to reduce CO_2 emissions and the relative inefficiency of the primary energy to electrical heating conversion).

Within the framework of a mandatory program, 6 million additional homes would contribute to the peak-demand-suppression system, as well as a number of the 10 million equivalent homes of the business sector.

If a single -1 °C demand-response cycle is allowed per day, each home (arbitrarily based on our sample home, and increasing the base power production for the reheat cycles

2001	2002	2003	2004	2005	2006	2007	2008	2009	2015	
79.6	79.7	83.5	81.4	86	86.3	59	84.4	91.5 (10/2009)	104	(forecast RTE 2009)

Figure 15.28   Peak consumption history in France and predictions (source RTE).

as explained above) can provide about 6 kWh over less than 3 h, within an hour or less, with a peak of about 3.75 kW. Together, the 6 million homes provide a fast capacity of over 30 GWh at a maximum power of 22 GW.

Most probably, the characteristics of our sample home do not reflect the average characteristics of the entire French residential market. Still, our evaluation gives us an order of magnitude that seems to clearly indicate that it is possible to completely replace the additional fast production capabilities, characterized by an extremely poor CO_2 performance for heating, by a voluntary expansion of demand response. The demand response is also more flexible than additional power plants, as it can provide additional power where it is required (e.g., in the often problematic western region of Brittany), without the constraints of the underlying transport network (limited incremental capacity during demand peaks, power losses).

Beyond the suppression of extreme peaks, there seems to be ample potential to also significantly smoothen the daily demand curve during winter: the portion of the demand beyond the daily average represents about 80 GWh, out of which about 20 GWh for the daily peak (Figure 15.27). From our evaluations above, it seems that the cumulated capacity of the 6 million homes would allow to shift about 30 GWh from peak to off-peak: enough to significantly flatten the daily demand curve, eliminating production and grid-capacity issues while decreasing the CO_2 emissions level and the average cost of electricity!

15.7.1.2   The Full Business Case of Demand Management

The business case associated to grid-balancing activities through demand shedding has been evaluated above. The yearly revenues associated to this activity are very dependent on the regulator decisions regarding the indemnification of additional costs of the electricity operator: while a standalone network balancing activity appears very profitable under the assumption that additional energy costs are considered a mutualized grid management cost, under any other assumption the yearly revenue per home is in the 10 to 20 € per year range, unlikely to justify the investments necessary for such an activity.

However, the potential of demand management goes beyond grid balancing. The cost of electricity is highly variable, depending on the time of the day (Figure 15.29). These varying prices simply reflect the market estimate of the instantaneous marginal production price of electricity.

As electricity operators consider that the demand side cannot be influenced, today's predominantly fixed electricity residential prices simply reflect the average of the price curves weighted by the consumption profile of the customer.

This situation creates a significant opportunity for demand management. For each MWh of consumption "moved" from the peak hours to the lowest-cost hours the gain ranges from 40 up to 90 € ! In the case of our sample home, a single demand shedding cycle of 1 °C shifts about 6.2 kWh. If this shift is optimized properly to the least-cost hours: the

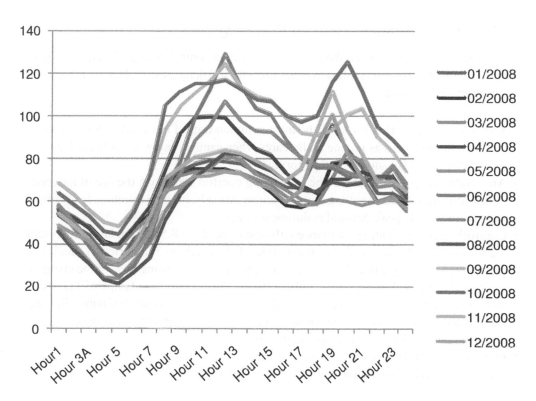

Figure 15.29   EPEX spot prices depending on delivery hour, monthly averages (2008).

resulting gain for the operator is in the 0.25 to 0.50 €per day. This represents 5 to 10% of the final consumer bill, as well as a significant marketing advantage : the CO_2 emission mix of the electricity sold is also improved. Of course, the same demand-response systems also introduce a better regulation of energy in the home, resulting in additional savings for the consumer.

The first operators to fully exploit these possibilities will be able to lower the electricity prices and will become more competitive.

Demand-response flexibility goes both ways: it is also possible to increase demand. This will make it easier for operators to source wind energy and other renewable-energy sources, as they will be able to temporarily "store" this randomly variable energy in the homes of their customers, in the form of hot water, slightly higher temperature, and so on.

We believe that the market dynamics will then lead to the general adoption of demand management, much in the same way as VoIP spread out from labs to every home in less than 10 years. In the same time frame, with the increasing penetration of electric cars, the potential of demand response – which will of course control the timing of battery recharging in homes and office parkings – becomes immense.

15.7.2 Demand Response Beyond Heating Systems

So far, we have only evaluated the potential of heating control during winter. Clearly the scope of demand response applies to other uses of electricity as well. Here is a list of the most promising domains:

- Sanitary hot water. This is one of the most common uses of electricity, and also one of the most flexible ones as no one really cares when sanitary hot water is heated, as long as one can take a hot shower in the morning.
- Air-conditioning systems. With the ever-rising comfort standards, the use of air conditioning during summer is getting more common. Already, in some countries for example, in Italy, the yearly peak demand is during summer.
- Controlling public lighting. In France only, there are about 8.5 million lamps representing 47% of the electricity bill of municipalities. Public lighting is notoriously inefficient, most lamps not only using obsolete technologies, but also wasting 30% of the energy to light up the sky, and adding to the light pollution. Replacing these lamps will take time and be very costly, however, a lot can be done to optimize the control of public lighting, such as using multilevel lighting (lower levels, e.g., one lamp out of two when there is no traffic and/or during peak power demand).
- Domestic and industrial refrigeration.
- Electric and hybrid car battery-recharging control.

15.7.2.1 Sanitary Hot Water

We consider a hot water tank of 200 l, with a daily hot water usage of 150 l. This corresponds to a daily water consumption of 500 l/day, with 50% warm water (warm water is cold water mixed with 60% of hot water from the hot-water tank).

In order to heat these 150 l from about 10 °C to about 60 °C, about 8.7 kWh are necessary each day. There are about 15% losses associated to thermal dissipation (the hot water temperature of sanitary hot-water tanks decreases by about 7 °C/day), therefore the total energy consumption will be about **10 kWh each day, representing about 3500 kWh/year**. It is generally considered that the energy consumption for sanitary hot water represents about 855 kWh per person per year, or about 25 kWh/m^2. This already represents about 10% of the total heating bill, but this share will increase in the future (see Figure 15.19).

The majority of sanitary hot-water tanks starts heating around 22:30, at the beginning of the reduced night tariff period. The heating period lasts about 4 h (4 × 2500 W = 10 kWh).

15.7.2.2 Ventilation

In France, active ventilation systems are mandatory in new buildings since 1983. Most systems are single-flow ventilation systems which simply pump air from the house without trying to recover thermal energy.

The fan engine has a power of about 50 W. An average ventilation system renews about 30% of the building atmosphere per hour (this is extremely variable in a range of 10% to 100%). For a 100 m^2 house, with a volume of 250 m^3, such an average ventilation system will pump out 75 m^3 per hour. If $T_{ref} - T_{ext} = 10$ °C, this represents a power loss of about 325 W (plus the engine power).

The yearly energy loss caused by these simple ventilation systems represent about 2400 kWh per year (1950 kWh of thermal losses, plus 450 kWh for the fan consumption).

During the public grid peak consumption periods, the outside temperature is below 0 °C: the energy gain by stopping the ventilation system is about 700 watts (650 thermal loss watts, plus 50 watts for the fan consumption).

15.8 Conclusion: The Business Case of Demand Response and Demand Shifting is a Key Driver for the Deployment of the Internet of Things

For years, electricity operators have considered that demand was a random process governed only by statistical laws. As a result, the electricity industry has developed extremely sophisticated strategies to adapt production to demand in real time. The costs associated to this lack of control on demand are very high:

- dimensioning of public grids for peak transmissions;
- building of "peak power" plants used only a few hours during the year;
- extremely inefficient CO_2 emissions during peak hours.

However, there was no easy alternative as electricity operators, despite early attempts with powerline communications, never really developed a communication network capable of monitoring and controlling demand at the consumer level in real time.

With the ubiquitous presence of the Internet, it seems clear to us that times have changed, and that the conditions are present for a fast deployment of demand-response technologies, which seem to be capable of bringing considerable flexibility to today's "statistical only" approach.

One of the interesting consequences of the development of demand response will be an increased level of competition among electricity operators.

Today, all operators work from the same statistical models of demand, and use the same power-plant technologies, facing identical costs. Competition is limited to limited marketing innovations and marginal production efficiency differences. If electricity operators were airlines, they would all be using the same planes, and there would be only one class of ticket: customers would choose according to seat color and meals.

After having deployed efficient demand-management tools, they will resemble more closely today's airlines, using sophisticated yield management, several classes of tickets, low cost or business class only strategies. For airlines, this has resulted in a better usage of airplane, air traffic route and airport capacity, resulting in dramatically lower tariffs.

For electricity operators too, the end result will be a maximization of the infrastructure potential, and as a consequence lower costs and lower emission levels.

However, this communication network and home controllers, do have a cost. Deploying such networks and controllers in a silo mode (the initial temptation of all utilities and TSOs) is clearly not the optimal model. Instead, these smart-grid applications should be considered as applications of the Internet of Things: the communication network and the home controllers should support general-purpose M2M communications and applications, decreasing the marginal cost (energy, amortization) of the smart-grid use cases. In order to enable this infrastructure sharing, standards such as IP and ETSI M2M are critical: IP ensures that the communication network is application and physical layer agnostic, ETSI M2M forms the middleware layer controlling the information flows: which application can access which sensor/actuator, for which usage.

16

Electric Vehicle Charging

16.1 Charging Standards Overview

Until 2010, the standardization of EV charging addressed only very basic aspects, such as how the vehicle should behave when it detects charging power or go back to sleeping mode when the charging power disappeared (IEC 61 681:2001). The plugs themselves conformed to national standards (SAE J1772 shown in Figure 16.3, VDE-AR-E 2623-2-2/Mennekes shown in Figure 16.1, JARI/TEPCO), or to IEC industrial connector specifications: IEC 62 196-1, single phase, for the US and Japan, and IEC 62 196-2 or 62 196-3 (or 60309, see Figure 16.2) for Europe and China.

However, the standardization efforts for EV charging have intensified since 2008, and address mainly the following domains:

- The **definition of a physical connector**. On the EV side, the US and Japan have converged to the J1772 2010 plug (EV plug), which is now used by automakers Chrysler, Ford, GM, Honda, Mitsubishi, Renault-Nissan (Figure 16.5), Tesla, Toyota. Virtually all vehicles in production or planned in 2011 will use this plug. The J1772 plug is suitable only for single-phase charging.

  In Europe in the beginning of March 2011 there was still some debate between a EV plug format proposed by German manufacturer Mennekes, and a format proposed by Italian manufacturer Scame. The additional complexity comes from the fact that three-phase power is much more common in Europe and actually usable in many homes for faster EV charging, and also from the various earth/ground connection regulations in each European country. The final consensus seems to be that there will be two standard EV side connectors:
- Type 1 connectors, which are simply J1772 single-phase connectors with 5 pins, a charging voltage up to 250 V and charging current up to 32 A, that is, AC charging power up to 7 kW (full charge in 4 to 6 h, or 50 to 70 km per hour of charge).

The Internet of Things: Key Applications and Protocols, First Edition.
Olivier Hersent, David Boswarthick and Omar Elloumi.

Figure 16.1   Mennekes EV plug (IEC 62196 v2 candidate).[1]

- Type 2 connectors, originally proposed by Mennekes, which enable single- or three-phase charging with a charging voltage up to 500 V, and a charging current up to 63 A (70 A for single-phase charging). Three-phase 400 V charging at 32 A represents a charging power of 22 kW (full charge in a couple hours, or 100 to 150 km per hour of charge).

In Europe, there is also a lot of activity regarding the definition of an EV charging equipment (EVSE) side standard plug format, which would introduce some decoupling with the EV side plug, as each car could use its own cable. As of May 2011, the proposal of the EVplug alliance seemed to have gained a wide consensus (Figure 16.4). This EVSE plug is also called the "type 3" plug.

At the international level, the current standardization converged so far only on the requirements (IEC 62 196-1:2003 Plugs, socket-outlets, vehicle couplers and vehicle inlets – Conductive charging of electric vehicles – Part 1: Charging of electric vehicles up to 250 A AC and 400 A DC).

Figure 16.2   IEC 60 309-2 3P+N+E, 9h plug.

---

[1] Courtesy of loremo http://www.flickr.com/photos/loremo/3499948469/.

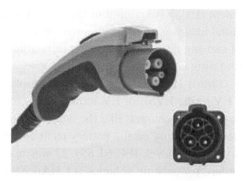

Figure 16.3 SAE J1772 plug.

- Enhanced security. Early EV chargers relied only on traditional 30 mA differential circuit breakers to detect current leakage. Modern EV chargers must have a dedicated pilot wire, which conforms to IEC 61 851:2010 (*Electric vehicle conductive charging system – Part 1: General requirements*), which implements proactive mass defect detection. This enhanced security mechanism is detailed in section "The IEC 61 851 pilot wire".
- Basic charging power control by the charging infrastructure. IEC 61 851:2010 also specifies a simple pulse width modulation (PWM) protocol on the pilot wire enabling

Figure 16.4 Type 3 plug on one of the first EVplug alliance charging station, deployed in France.

the EV charging equipment (EVSE) to communicate to the EV the maximum charging current allowed. The initial idea was to make sure that the EV would never draw excessive currents capable of causing fires in the charging infrastructure, but the mechanism can also be used for admission control and demand response.

- Bidirectional charging control (DC charging). This functional area covers also the communication by the EV to the EVSE of EV requirements related to the charging of the EV battery by an external CC charger: like the constant current, constant power or constant tension required (depending on the battery technology). There are no agreed international standards yet at this level, IEC 61 851-23 was not published as of March 2011. The *de-facto* standard for existing vehicles is CHAdeMO, which uses dedicated pins on the J1772 connector.
- High-level communication. This functional area encompasses the requirements for charging control, but also covers security, charging and potentially many other services. At the international level, there are two standardization tracks: ZigBee SEP 2.0 (see Chapter 13), and IEC 15 118 (*Road vehicles - Vehicle to grid communication interface*). At the physical level, the proposals are to use IPv6 on top of Homeplug Greenphy, G3 translated to the 150–450 kHz frequency band (300 to 400 kbit/s), or CAN (CHAdeMO proposal).

16.1.1   IEC Standards Related to EV Charging

The main IEC standard related to EV charging is IEC 61 851, managed by IEC TC 69. IEC 61 851 is split into several documents:

- **Part 1**: General requirements;
- **Part 21**: Electric vehicle requirements;
- **Part 22**: AC charging station requirements;
- **Part 23**: DC charging station requirements;
- **Part 24**: Communication protocol.

The IEC documents 3 cases for the physical connection:

- **Case "A"**: the cable and plug are attached to the EV.
- **Case "B"**, the cable is detachable. In case B1 the cable connects to a standard domestic plug, in case B2 the cable connects to a specific charging station. In practice, EVs sold in 2011 usually include a "B1" cable by default.
- **Case "C"**, the cable is attached to the EVSE.

The IEC also defines several charging modes. This nomenclature is now used by all EVSE vendors:

– **Mode 1** uses a standard 16 A socket outlet, single phase or 3-phase. The protection against electric shock relies only on differential breakers in the charging infrastructure or in the cable. This charging mode is prohibited in the US, but outside of the US most EVs currently circulating use mode 1 charging, and mode 1 charging is still proposed by default in many new EVs in 2011.

– **Mode 2** still uses a standard 16 A or 32 A socket outlet on the charging infrastructure side, but uses a specific plug on the EV side. The cable must include a protection device that implements proactive current-leakage detection through the pilot wire. The additional PWM communication over the pilot wire remains optional. This mode is the default for newer EVs in the US and Japan (the J1772 plug is used on the EV side), and is available as an optional cable for most recent EVs in other countries.

– **Mode 3** requires a dedicated EVSE. The pilot wire that controls the charging session is managed by the EVSE. No charging power is available over the charging cable until the EV has been detected by the pilot wire and the absence of current leaks or ground defects has been verified. The pilot wire also optionally indicates the maximum charge level by using PWM modulation: 16 A (normal), 32 A (semifast), > 32 A (fast). Mode 3 is safer than mode 1 or 2, because the plug is energized only if all of the following conditions are met:

  • the vehicle power plug is totally inserted (the pilot pin is last to connect);
  • ground continuity has been checked (pilot current present);
  • the vehicle has transmitted a signal confirming that everything is secured and charging is ready to begin.

– **Mode 4** applies to external chargers (e.g., DC charging) and also covers the inverse mode where the EV provides power to the charging infrastructure. This mode, covered by IEC 61 851-23, was not fully specified as of March 2011. It requires a serial bidirectional communication over the pilot wire.

In practice, the new EVs sold in 2011 usually had two plugs, a standard domestic male plug for mode 1, and a J1772 or equivalent plug for modes 2 and 3 charging (Figure 16.5).

16.1.1.1   The IEC 61 851 Pilot Wire

IEC 61 851:2001 requires for mode 2, 3 and 4 charging a dedicated "pilot wire" in the EV plug. This pilot wire is connected by a 1000 Ohm resistor to the mass (ground) of the vehicle and a weak current circulates from the pilot wire to the ground of the connector in order to test the correct grounding of the EV.

The new IEC 61 851-1 (edition 2 of November 2010) introduces an additional control function, and documents in annexes A and C a (non-normative) PWM protocol for the pilot wire (Figure 16.6).

In Figure 16.6, the EV indicates that it is ready to begin a charge session by closing switch S2, which changes the voltage measured at point Va. On the EVSE side, the duty

Figure 16.5   J1772 (type 1) connector on a Renault EV.

cycle of signal Vg indicates the maximum current that the EV is allowed to draw from the EVSE at any moment: no charging for a duty cycle lower than 3%, and then any value from 6 to 80 A coded by a duty cycle above 8%. A special value of the duty cycle of 5% indicates that a digital communication ("high-level communication") is used instead.

The standard indicates that the dedicated pilot function could be replaced by a current modulation on the ground conductor (61851-1ed2, Appendix C), without defining it yet.

In this new version, the mandatory EVSE/charging cable functions for charging modes 2, 3 and 4 are:

– the verification that the EV is connected;
– continuous control of the correct grounding of the EV;
– energization and de-energization of the EV.

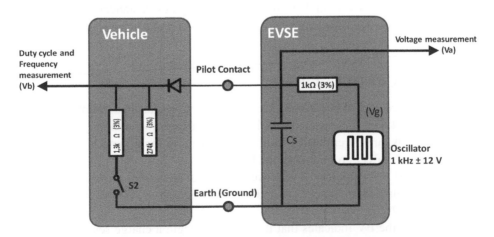

Figure 16.6   Pilot wire circuit, according to IEC 61851 and J1772.

The new optional functions are:

- selection of the charging power;
- determination of the ventilation requirements of the charging area;
- bidirectional power control (not yet specified in edition 2 of IEC 61851).

16.1.1.2 High-Level Communication: IEC 15 118

IEC 15 118 *Road vehicles - Vehicle to grid communication interface* was still a work in progress at the time of writing. This standard is managed by ISO/TC 22 (Road vehicles) jointly with IEC TC69.

- IEC 15 118-1 "General information and use-case definition" defines the vocabulary used in other parts of the standard, and focuses on the use cases for high-level communications between the electric vehicle communication controller (EVCC) and the supply equipment communication controller (SECC). Both online (Figure 16.8, SECC connected to the E-Mobility operator by a communication network as the service is provided) and semi-online (Figure 16.7, SECC-E-mobility provider communication is

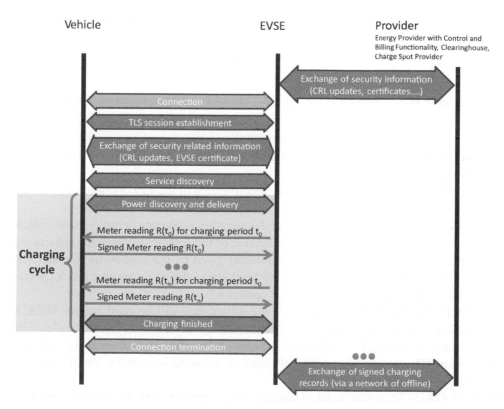

Figure 16.7 Overview of IEC 15 118-2 charging session semionline protocol flows.

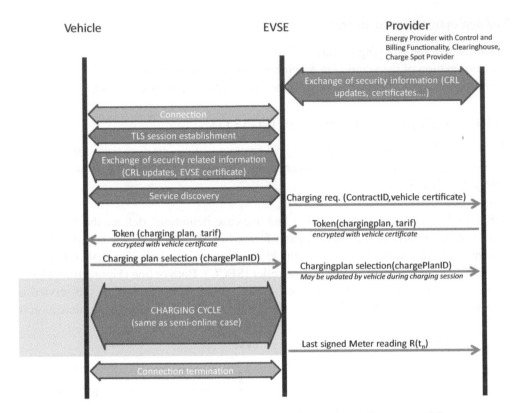

Figure 16.8   Overview of IEC 15 118-2 charging session online protocol flows.

not synchronous with the service provided) use cases are defined. IEC 15 118-1 defines four identifiers:

- **SECC Spot Operator ID:** the unique identification of the spot operator that provides the energy to the vehicle.
- **SECC Power Outlet ID:** the unique identification of the power outlet to the vehicle.
- EVCC Provider ID: the unique identification of the contract between the vehicle user or the vehicle itself and the energy provider given by the E-mobility operator. The E-mobility operator is defined as "the legal entity that the customer has a contract with for all services related to the EV operation".
- **EVCC Contract ID:** it identifies the contract that will be used by the SECC to enable charging and related services (incl. billing). It is associated with the electricity consumer (who can be the driver, the owner of the vehicle or a E-mobility operator).
- **Session ID:** the identifier of the charging or value added service session, obtained after the authentication procedure.
- IEC 15 118-2 "Technical protocol description and open systems interconnections (OSI) requirements", defines an application layer protocol on top of IPv6 (optionally IPv4) between the SECC and the EVCC. The communication uses a TCP connection secured

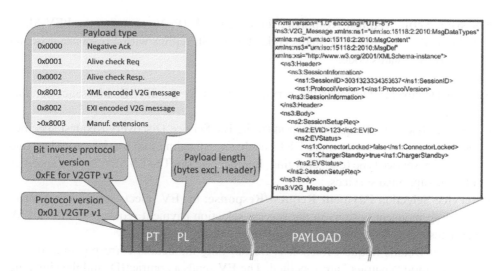

Figure 16.9 V2GTP header structure and example XML payload.

by TLS, the SECC acting as the TLS server, and uses the vehicle to grid transfer protocol (V2GTP) defined in IEC 15 118-2 (see below for more details on V2GTP). the SECC discovery protocol (SDP)

• IEC 15 118-3 "Physical and data link layer requirements", will define the physical layer supporting the IP connectivity for the V2G protocol. It was not stable at the time of writing this document, although it seemed that PLC HomePlug GreenPhy technology (refer to the PLC chapter) had the best chance of being selected.

The V2GTP session is bootstrapped using the SECC discovery protocol (SDP). The EV begins by multicasting a UDP_SECC_LISTEN message to port 15 118 (!) as a means of discovering the SECC. The EVSe generates a random sessionID or provides the last valid sessionID if this is an attempt to resume a session. The SECC, if any, will respond to the source UDP port with a SECC discover response message, which specifies the server-side TCP port for the V2GTP session.

As V2GTP uses a TCP connection (which is stream oriented) to transport messages, it needs a framing mechanism. The V2GTP header (Figure 16.9) serves this purpose.

V2GTP defines an XML message set, which can be encrypted and signed using a certificate bound to a given contractID (current status at the time of writing). All messages have a common header format that carries the session ID.

The V2G protocol session is organized as follows:

Initialization of communication session: after session bootstrapping with the SECC discovery protocol and TLS session setup, the following messages are exchanged:

• **SupportedAppProtocolRequest/Response:** EVSE and EVCC exchange-supported protocol major/minor versions, associated namespaces and preference order.

- **Session Setup Request/Response:** the plug-in EV (PEV) specifies the PEV MAC address (PEVID) and PEV status code. The EVSE replies with a response code, an EVSEID and status code.

Service Discovery: Discover the services offered by the EVSE. Messages for this activity:
- **Service Discovery Request/Response:** the EV Service Discovery Request specifies the service scope (one or more URIs, each corresponding to a service provider), and optionally service type (e.g., charging, Internet access). The EVSE response specifies a list of supported services.
- **Service Selection Payment Request/Response:** the EV selects services and corresponding payment options in the service selection payment request, and the EVSE validates each service/payment option selected.
- Payment Details Request/Response: So far, only the details corresponding to payment method "contract" are specified. The EV sends a contractID, and the signature certificate chain of the EV. The EVSE responds with a random challenge that has to be signed by the EV.
- **Contract Authentication Request:** the EV sends a copy of the challenge and its signature, and the EVSE answers with a response code.

Charge vehicle: charging the EV is one possible service offered by the EVSE. It is divided into three phases:
(1) **Set up charging process:**
   - **Charge Parameter Discovery Request/Response:** the PEV provides status information, charging mode (DC or AC), desired point in time for the end of charge, estimation of required recharge energy, maximum charge power, maximum charge voltage, minimum charge voltage, maximum number of charge phases, maximum charge current and minimum charge current. The EVSE responds with its status information and provides its own charging-limit parameters, as well as the identity of the relevant mobility/energy provider and a tariff table (Figure 16.10).
   - **Line Lock Request/Response:** EV requests the EVSE to lock the connector on the EVSE side and specifies its own lock status.
   - **Power Delivery Request/Response:** the PEV requests the EVSE to start providing charging power, specifies the selected tariff and optionally transmits the estimated charging profile (Figure 16.11) it will follow during the charging process. The EVSE replies with a response code.

Figure 16.10   TariffTable structure IEC15118 draft (may 2011).

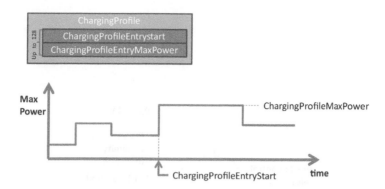

Figure 16.11 Charging profile structure in current IEC 15 118 draft (May 2011).

- **Metering Status Request/Response:** the EV request the EVSE to send its current metering status. The response of the EVSE contains a timestamp, an instantaneous estimate of the power the EVSE could deliver, optionally the current power it is delivering, a meterID and current metering information.
- **Metering Receipt Request/Response:** the EV Metering Receipt Request contains the electronic signature of the following information elements: sessionID, MeterInfo, PEVstatus, time stamp, tariffID, metering information. This information ensures nonrepudiation for the billing process of the E-mobility provider. The EVSE acknowledges with a response code.

(2) **Charging process:** periodic metering status request/response and metering receipt request/response (the draft standard mentions a periodicity of about 10 s), and if needed charge parameter discovery request/response and power delivery request/response.

(3) **Finalize charging process:** stop charging process (final power delivery request/response, metering status request/response, metering receipt request/response), unlock charge cord with line lock request/response.

At the time of writing (May 2011), there were some discussion ongoing as to whether to continue investigating a separate XML message set, or to use the ZigBee SEP 2.0. The openV2G project (http://openv2g.sourceforge.net/) is implementing both solutions. One of the issues is that the current design of V2GTP follows the style that was used in the 1990s for telecom protocols (e.g., H.323 for VoIP), and is not aligned with the resource-based (REST) programming style used by most recent protocols (ZigBee SE2.0, ETSI M2M . . .). This means that, as outlined in the current draft, V2GTP will not be able to leverage modern protocols designed for REST applications, like CoAP or HTTP.

16.1.2 SAE Standards

16.1.2.1 J1772 "Communications Between Plug-In Vehicles and the Utility Grid"

The original J1772 plug format was designed in 1996 by Japanese manufacturer Yazaki, and adopted by the SAE hybrid standard committee in 2001. This document defines the

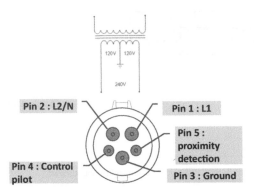

Figure 16.12   J1772 plug, and typical US residential wiring.

physical format of the EV plug for AC charging and addresses basic security and power control requirements by means of a proximity detector pin and a control pilot pin.

The original J1772:2001 specification used 9 pins, plus a proximity detector. 2 pins were used for DC charging up to 400 A, 2 pins for AC charging up to 40 A, 1 ground pin, 1 control pilot pin, and 3 pins for SAE J1850 data transmission. A magnetic proximity detector switch was provided on the EV side to prevent motion during charging.

J1772 was revised on January 2010. The J1772:2010 connector, illustrated on Figure 16.12, only uses 5 pins: two pins for single phase AC charging up to 80 A, one ground pin, one pilot wire pin, and one proximity detector pin. The proximity detector pin is connected to the ground via a resistor network in the plug, the resistor value changes when the latch release switch of the plug is pressed, signaling that the plug is about to get disconnected. The control pilot protocol is identical to that of IEC. The proximity detector and pilot wire pins are shorter, so that they connect last and disconnect first. The 3 dedicated pins provided by J1772:2001 were removed as SAE now plans to use PLC for high-level communications.

J1772 defines 3 charge levels:

- **"Level 1"**: charge via a charge cable including the pilot wire control equipment ("mobile EVSE"), connected to a US domestic plug (single-phase 120 V/15–20 A).
- **"Level 2"**: charging via a fixed EVSE station connected to a 240 V single phase, up to 80 A.
- **"DC charging"**, using an off-board charger.

16.1.3   J2293

Task force J2293 of the SAE is focused on the standardization of EVs and the communication between the EV and the EVSE. It initially addressed both conductive and inductive charging, but inductive charging (J1773) is no longer considered. The J1772 standard, which focuses on conductive charging, implements the same pilot wire mechanism as the IEC.

The 2008 version of J2293[2] described a high-level charging protocol based on the J1850 serial communication protocol, which has never really been used and was formally made obsolete in 2010.

The SAE worked on a comprehensive list of use cases for EV charging, published as J2836/1 (a 244 page document!) in April 2010.

The SAE then worked on a revised architecture and communication protocol supporting all these use cases, and considered two options for the high-level communication protocol:

- The CAN bus that is adopted is all modern vehicles, but would need a specific message set to support EV charging.
- IPv6 and the ZigBee SEP 2.0 REST data model. Initially both 6LoWPAN over 802.15.4 and IPv6 over CPL were considered, at present it seems the SAE is focusing on IPv6 over CPL, most likely HomePlug GreenPhy (see Chapter 2).

This revised architecture was published in June 2010 as J2847-1 *"Communication between the Plug-in vehicle and the utility grid"* and J2847-2 *"Communication between Plug-in Vehicles and the Supply"*(which adds a message set to support DC charging and replaces J2293). Both documents are based on ZigBee SEP 2.0 (see Chapter 13).

16.1.4   CAN – Bus

The CAN bus is a proprietary protocol designed by Bosch GmbH in 1983, which was then published at SAE in 1986. CAN 2.0 was published in 1991, and uses an expanded 29-bit identifier format.

CAN is a half-duplex multimaster broadcast serial bus using NRZ encoding. Originally, only the interface to the physical layer was specified, but most recent CAN implementations rely on the physical layer specified by ISO 11 898-2:2003 (*"Road vehicles -Controller area network (CAN) - Part 2: High-speed medium access unit"*) or ISO 11 898-5:2007 (*"Road vehicles - Controller area network (CAN) - Part 5: High-speed medium access unit with low-power mode"*). The achievable bitrate varies between 1 Mbit/s (distances up to 40 m with ISO 11 898-2 or -5) and 125 kbit/s (distance up to 500 m with ISO 11 898-3).

The CAN data link layer uses ISO 11 898-1:2003 ("Road vehicles – Controller area network (CAN) – Part 1: Data link layer and physical signaling"), which specifies the medium access control sublayer and the logical link control sublayer. The MAC sublayer implements a CSMA collision avoidance mechanism: all messages begin with an ID,

---

[2] J2293-1 "Energy Transfer system for Electric Vehicles – part 1: Functional Requirements and system Architectures", and J2293-2 "Energy Transfer system for Electric Vehicles – part 2: Communication Requirements and Network Architecture".

which also encodes the priority level: in case a "1" and a "0" are simultaneously transmitted on the bus, only value "0" will be sensed by all nodes in the network. The node that had just transmitted a "1" then knows a collision has occurred and stops transmitting. The original CAN and CAN 2.0A specification uses an 11-bit identifier field. On recent CAN 2.0 implementations using a 29-bit identifier structure: the first 3 bits are reserved for the priority level, the next 18 bits encode the frame format (broadcast or unicast parameter group, and 8-bit destination address/group), the last 8 bits encode the sender address.

The ID field is followed by up to 8 payload bytes. Any higher-layer protocol may be used on top of CAN, for instance:

- The ISO transport protocol (ISO TP, 15 765-2) is used by existing diagnostics protocols on CAN (e.g., ISO 14230), and provides segmentation (transmission of up to 4095 bytes), flow control, broadcast and unicast addressing. Error recovery is left to the application layer.
- TP 2.0 that adds automatic error recovery (mainly used by Volkswagen).
- CAN Open, developed by Bosch GmbH and handed over to CAN in Automation user group (http://www.can-cia.org/) in 1995. It was standardized as EN 50 235-4 in 2002, which defines a object dictionary enabling cross-manufacturer compatibility.
- SAE J1939, which defines a segmentation and flow control mechanism, and an application layer for buses, trucks, and agricultural equipment.

At present, there is no proposal to port 6LoWPAN over CAN, which means the CAN option for EV to EVSE communication would not be IP based and would serve only for low-level control purposes. This is considered problematic by some car manufacturers who believe the EV to EVSE connection may be leveraged also for other uses, for example, downloading movies while charging, and in general would require the versatility of IP.

CAN, which was designed for closed systems, also lacks a set of specifications for strong authentication and security, which are required for communications between the EV and EVSE.

16.1.5 J2847: The New "Recommended Practice" for High-Level Communication Leveraging the ZigBee Smart Energy Profile 2.0

The J2847 series is the result of joint work since 2009 between SAE and the ZigBee and HomePlug alliances, with the following objectives:

- defining a message set that would support the requirements of J2836/1;
- integrate with home area networks and utility networks;
- follow the REST model recommended by the NIST, and be based on IPv6.

Because the high-level communication is based on IPv6, any underlying transport layer supporting IPv6 might be used. The REST interactions might be transported by HTTP over IPv6 for high-speed physical layers such as HomePlug GreenPhy or G3, or by CoAP over 6LoWPAN for lower-speed physical layers such as 802.15.4 or low-energy CPL. Refer to Chapter 2 for more details on CPL technologies.

The original ZigBee SEP 2.0 specification does not address the EV charging use case only, therefore the SAE reworked the specification and split it in several documents focused on the specific requirements for EV charging:

- **J2847/Part 1** "Communication between Plug-in Vehicles and the Utility grid" lists the messages used to support the J2836/1 baseline requirements:
  - vehicle, customer, evse identification and authentication;
  - energy request management (energy and power requests, energy and power management and scheduling);
  - timing information (start and stop time, anticipated duration, charging profile, actual start time);
  - pricing, including time of use pricing, critical peak pricing;
  - load control (demand response, load shifting);
  - vehicle information and status.

  The other J2847 documents were not published at the time of writing:

- **J2847/Part 2** *"Communication between Plug-in Vehicles and the Supply"* will replace the high-level messages specified in J2293 for the specific use case of DC charging.
- **J2847/Part 3** will focus on the *Reverse Power Flow* (RPF) use case, that is, when the EV provides energy to the utility network.
- **J2847/Part 4** will focus on charging system and EV diagnostics.
- **J2847/Part 5** will focus on vendor-specific extensions and options.

16.2 Use Cases

16.2.1 Basic Use Cases

The basic use case is, of course, the charging of the EV battery. The battery of a plug-in hybrid electric vehicle (PHEV) has a capacity of typically 5 to 6 times the capacity of a hybrid electric vehicle (HEV). The capacity of the battery of pure battery electric vehicles (BEV) is 3 times or more that of PHEVs (see Figure 16.13).

In order to allow for reasonable charging times, the charging power and current is typically 16 A or more. For an isolated home, this is enough to require load management and limiting in order to avoid simultaneous use of other high-power appliances, such as electric heating and sanitary water, electric ovens, and so on. In a residential complex, the simultaneous charging of several plug-in vehicles will also require careful synchronization of management in order to ensure fair charging of all vehicles, without exceeding the power limits of the connection between the building and the utility grid.

In both cases, this means that charging stations at home or in the work place will either require a separate utility connection (as in the case of Italy where most existing residential

		PHEV15	PHEV40	BEV
Battery capacity (kWh)		5	16	24
Charging time (20% to 100%)	1.4 kW (16 AWG)	2 h 50 mn	9 h 10 mn	13 h 45 mn
	3.3 kW (16 AWG)	1 h 15 mn	3 h 50 mn	5 h 50 mn
	7 kW (8 AWG)	35 mn	1 h 50 mn	2 h 45 mn
	19.2 kW (8 AWG)	15 mn	40 mn	1 h
	60 kW (4 AWG)	5 mn	15 mn	20 mn
Charge-depleting range or All Electric mode range		23 km−15 miles	64 km−40 miles	160 km−100 miles

Figure 16.13   Typical battery capacity and charging time of EVs.

electric connections cannot deliver more than 3 kW), or will need to be integrated in the local energy-management system. The latter is expected in all countries where the cost of upgrading the capacity of the existing home connection is lower than the cost of a new connection to the utility grid (e.g., in France where a new connection is charged about 1000€ in the best cases, and much more depending on the distance between the home and the closest distribution operator street cabinet).

Considering only the short term, EVs are beneficial for the stability of the electric grid. All transmission system operators (TSOs) face short-term mismatch problems between power production and demand (see Section 14.4 for more details). The charging power of EVs can be adapted in real time by using the pilot wire, and provide an ideal flexible load for demand-response policies.

Unfortunately, the EV is also problematic for utilities: EVs are ideal vehicles for commuters who run short distances between their work place and home. This means that the charging load will exhibit a pendular behavior as well: large and sudden charging requirement peaks in the suburbs after work hours and in business districts in the morning. However, since the midterm (hourly) pattern of such demand is predictable, utilities with smart-grid capabilities may implement load-shifting and load-capping policies to smooth the total demand in a given district: for instance the HVAC installations of business centers may be programmed to reach their setpoints just before the rush hour, possibly to slightly exceed the heating setpoint during winter and cooling setpoint during winter so that additional energy can be delivered to the EVs during a couple hours.

The basic capability provided by the pilot wire PWM charging rate control is sufficient to implement load limiting, load shifting and demand response, if all vehicles are treated the same, and of course in the specific case of a single home with a single car.

However, not all vehicles have the same needs:

– The battery of some vehicles may be charged at 90% of capacity already, while other cars may be at a critical battery level.
– Some vehicles may need to depart within the next hour, while other vehicles will not move for 8 h or more.
– Some users may be willing to pay more to be served in priority.

These examples show that more sophisticated arbitration policies that seek to optimize the aggregate satisfaction of users need more information on each EV state and requirements: the high-level communication protocol, such as ZigBee SEP 2.0/J2847, will become a mandatory feature as soon as EVs reach a penetration that make it very likely to have 5 EVs or more (including rechargeable hybrids) simultaneously connected in an office or residential complex parking, this is probably as soon as 2015!

16.2.2 A More Complex Use Case: Thermal Preconditioning of the Car

Traditional thermal engines have a nice side effect: heating is provided "for free". However during hot summer days, we have all noticed how air conditioning affects the car performance and consumption.

This air-conditioning tax will be a burden for EVs, both during summer and winter. As much as 30% of the total energy of the battery may be spent by air conditioning, and much of it during the first 10 min for the initial conditioning: typically 6 kW for 10 min and 2 kW afterwards for heating, 3 to 4 kW initially then 2 kW for cooling.

In order to minimize this problem, a possible strategy is to precondition the car. A study published in November. 2010 by the US National Renewable Laboratory (NREL)[3] quantified the improvement of preconditioning for several types of cars. The results provided in Figure 16.14 assume a mixed drive cycle of 55% city driving, and 45% highway driving. The impact of preheating would be even more favorable for short commuting cycles.

Of course, in addition to the increased range and reduced fuel consumption, the level of comfort is also improved by preconditioning.

This more complex use case shows that the communication protocols for future charging systems may need additional parameters, such as:

– The estimate of the preconditioning energy and power schedule, including the initial high-power conditioning and the ongoing conditioning. This should be calculated by the car based on external temperature measurement and air-conditioning setpoints.
– The preconditioning schedule, for example, when preconditioning is scheduled to begin.
– Whether or not the user charging profile allows preconditioning.

---

[3] "Analysis of Off-Board Powered Thermal Preconditioning in Electric Drive Vehicles" available at http://www.nrel.gov/vehiclesandfuels/vsa/pdfs/49252.pdf.

	Charge-depleting (CD) or All-electric range (AER) reduction (additional fuel consumption)		Range increase by 20 mn preconditioning (compared to no pre conditioning)	
Car category	Heating (−6°C ambient temperature)	Cooling (+35°C ambient temperature)	Heating (−6°C ambient temperature)	Cooling (+35°C ambient temperature)
100 mile BEV (e.g., Nissan Leaf)	−34%	−32%	+4%	+2%
40 miles PHEV (e.g., GM Volt)	−35% (+60% fuel)	−34% (+57% fuel)	+6% (−3% fuel)	+4% (−1%)
15 mile PHEV (e.g., Toyota Prius)	−19% (+3% fuel)	−32% (+49%)	+19% (−1% fuel)	+5% (−1% fuel)

Figure 16.14   Effect of air conditioning on range (55% city, 45% highway driving), from NREL study.

16.3   Conclusion

The need for communication systems in the car goes much beyond EV charging. The potential for M2M service providers is enormous: there are about 50 million cars produced in the world each year, for a total of over 600 million in circulation.

The European eCall directive, which mandates installation of automatic emergency calling in new cars as of 2014, has made it mandatory to install a cellular communication platform in all new European cars. Building on this, most automakers are now planning to use this cellular communication platform to implement a M2M GPRS connection for maintenance and monitoring of the vehicle. Such systems are already in place in all new EVs and will soon generalize to all new cars.

What is the next step? Automakers will soon provide on-board IP connectivity for the car appliances (GPS systems for instance). It would make no sense to build separate vertical systems for EV charging, car monitoring, and on-board M2M connectivity. Clearly, our vehicles will need to provide a generic M2M infrastructure able to leverage both the cellular network and IP connectivity provided through charging stations (or, in the future, by parking lots). With this in mind, we believe that the protocols as defined currently will need to evolve:

- IEC15118 current draft is "too vertical" but explores in details all the semantics for EV charging in the context of a multiactor deployment.
- ZigBee SE 2.0 is REST-based and a little more generic, but still focused on energy management. Multiactor interactions are not as clearly defined as in IEC 15 118.

ETSI M2M (see the ETSI M2M chapter) provides a generic REST framework for M2M, but so far no detailed work has been done to explore how the semantics of IEC15118 and SE2.0 could be integrated in this framework. The authors believe, however, that this is a very promising direction. The future communication architecture of our cars probably will need the use case versatility of frameworks such as ETSI M2M, complemented by specific interworking profiles for charging specific "REST resources" like those defined by IEC15118 and SE2.0. If that happens, EV charging "protocol" specifications should evolve into "REST resources" specifications, leaving the plumbing to other, more generic, protocols.

Appendix A

Normal Aggregate Power Demand of a Set of Identical Heating Systems with Hysteresis

For simplification, we suppose that weather conditions remain constant and that the temperature evolution curve in all homes is a zig-zag function of time (Figure A.1): when the heating system is on, the temperature increases by X °C/hour, when it is off, it decreases by Y °C per hour. All homes are equipped with a traditional thermostat that regulates the temperature with a hysteresis of H °C: the heating system is switched on when the temperature reaches T_{ref}, and switched off when it reaches $T_{ref} + H$. In reality, these temperature evolution curves are exponential functions of time, but this linear approximation is valid if H is small compared to the $T_{ref}-T_{ext}$, the temperature gradient between the inside and the outside of the house walls. The heating system therefore remains switched on for H/X hours (the part of the zig-zag during which temperature increases), and then remains switched off for H/Y hours (the part of the zig-zag during which temperature decreases).

We further suppose that initially thermostats are not correlated: at any moment, if we sort homes according to the last time the heating system was switched on, an identical number $k.\delta t$ of homes had their heating system last switched on between T and $T + \delta t$. Note that the last time the heating system was switched on at any time T must be within interval $[T–H/X–H/Y, T]$, therefore we can write the total number of homes as $N = k(H/X + H/Y)$.

Let us calculate the number of heating systems that are switched on at any point in time T: the moment they have been last switched on must be in interval $[T–H/X, T]$, therefore we have $k.H/X$ heating systems switched on.

The Internet of Things: Key Applications and Protocols, First Edition.
Olivier Hersent, David Boswarthick and Omar Elloumi.
© 2012 John Wiley & Sons, Ltd. Published 2012 by John Wiley & Sons, Ltd.

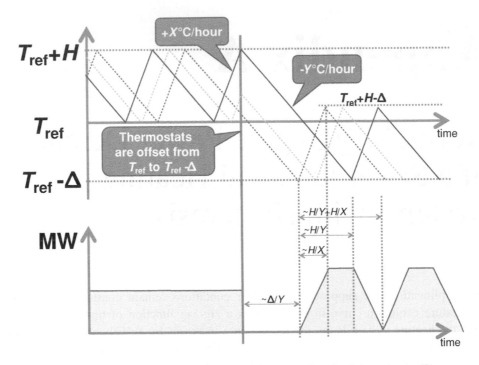

Figure A.1   Evolution of power demand after a synchronized thermostat offset.

If we call P_{max} **the cumulative power demand of all homes if they all had they heating systems switched on**, the average aggregate power demand of a set of homes with decorrelated thermostats is $P = P_{max}(H/X)/(H/X + H/Y) = P_{max}Y/(X + Y)$.

> ## Normal aggregate power demand of a set of homes with decorrelated thermostats:
>
> $$P = P_{max}Y/(X + Y).$$

Appendix B

Effect of a Decrease of T_{ref}. The Danger of Correlation

If we decrease T_{ref} by Δ °C at time 0, the effect on power demand depends on our thermostat strategy:

- We may switch off all heating systems (as by definition, all homes have a current temperature greater than T_{ref}, therefore also greater than $T_{ref} - \Delta$): the aggregate power demand of our homes instantaneously falls to zero.
- As the new hysteresis settings of our thermostat is now $[T_{ref} - \Delta, T_{ref} + H - \Delta]$, if $H > \Delta$ some homes have a current temperature in interval $[T_{ref}, T_{ref} + H - \Delta]$. We can choose to heat some or all of these homes up to $T_{ref} + H - \Delta$, maintaining residual power demand.

Let's choose the first strategy and evaluate its effects:

- Power demand will fall to 0 for Δ/Y hours: this duration represents the time it takes for the homes that were coldest (at T_{ref}) when the thermostat setting was changed to reach $T_{ref} - \Delta$.
- After Δ/H hours, heating systems are switched back on in an increasing number of homes. At time $\Delta/H + t$, the heating systems of all homes that reached temperature $T_{ref} - \Delta$ between time Δ/H and time $\Delta/H + t$ are switched on, that is, homes that had a temperature between T_{ref} and $T_{ref} + Yt (t < H/Y)$ when we changed the thermostat settings. These homes fall into two categories:
  - Homes that were in "cooling down" mode, the heating systems of which were last turned on in the interval $[-H/X - H/Y, -H/X - H/Y + t]$: there are kt such homes.

The Internet of Things: Key Applications and Protocols, First Edition.
Olivier Hersent, David Boswarthick and Omar Elloumi.
© 2012 John Wiley & Sons, Ltd. Published 2012 by John Wiley & Sons, Ltd.

- Homes where the heating systems were on, and were turned off at time 0. The heating systems of these homes were last turned on in the interval $[-Yt/X, 0]$: there are kYt/X such homes.

Overall, at time $\Delta/H + t$, $kt(1 + Y/X)$ homes have had their heating system turned back on, and at time $\Delta/H + H/Y$, all homes have had their heating systems turned back on. But in the meantime, at time $\Delta/H + H/X$, the first homes that had their heating system turned back on reach temperature $T_{ref} + H - \Delta$ and get turned off: the number of active heating systems stabilizes to $kH/X \times (1 + Y/X)$, and power demand stabilizes to $P_{max} \times Y(1 + Y/X)/(X + Y) = P(1 + Y/X)$.

At time $\Delta/H + H/Y$, the heating system of the last homes to reach $T_{ref} - \Delta$ is turned back on. After this time, this additional demand therefore stops, while $k(1 + Y/X)$ heating systems per unit of time are still being switched off, having reached $T_{ref} + H - \Delta$: power demand decreases again, and zeroes at $t = \Delta/H + H/Y + H/X$.

At this point, the first homes that reached $T_{ref} + H - \Delta$ have had time to cool off, and reach $T_{ref} - \Delta$. $k(1 + Y/X)$ heating systems per unit of time are being switched on, and energy demand increases again. This scenario is illustrated in Figure A.1.

> After an initial period where energy demand decreased to zero, energy demand becomes periodical, because we created a synchronizing event for all previously decorrelated heating systems.

Appendix C

Changing T_{ref} without Introducing Correlation

It is possible to decrease T_{ref} without creating a synchronizing event: for instance at $t = 0$ we can start to change T_{ref} to $T_{ref} - \Delta$ in each home, *but only when the home temperature reaches T_{ref}* (Figure C.1).

With this new strategy, the effect on power demand is progressive: for each unit of time the heating system in k homes reaching temperature T_{ref} remains off: aggregate energy demand decreases from $k.H/X$ to zero in H/X hours. After Δ/Y h, the first homes reach $T_{ref} - \Delta$, the thermostat turns the heating system back on, and aggregate energy demand increases again, before stabilizing to the same level as before we decreased the thermostat temperature.

C.1 Effect of an Increase of T_{ref}

Homes participating in a demand-response program accept only temporary changes to their home's temperature. At some point therefore the thermostat temperature must be increased again from $T_{ref} - \Delta$ to T_{ref}.

In order to avoid creating any synchronizing event, a possible strategy is to change the thermostat temperature for all homes that reach $T_{ref} + H - \Delta$ (Figure C.2).

If $\Delta/X < H/Y$ (case of (Figure C.1), the number of active heating systems increases from $k.H/X$ to $k.(H + \Delta)/X$ in Δ/X hours: energy demand increases to $P(H + \Delta)/H$, then remains constant, above average, for $H/Y - \Delta/X$ hours, then returns to its average value in Δ/X hours.

If $\Delta/X > H/Y$, the number of active heating systems increases from $k.H/X$ to $k.(H/X + H/Y)$ in H/Y hours: energy demand increases to P_{max}, then remains constant at the maximum possible value P_{max}, for $\Delta/X - H/Y$ hours, then returns to its average value in H/Y hours.

The Internet of Things: Key Applications and Protocols, First Edition.
Olivier Hersent, David Boswarthick and Omar Elloumi.
© 2012 John Wiley & Sons, Ltd. Published 2012 by John Wiley & Sons, Ltd.

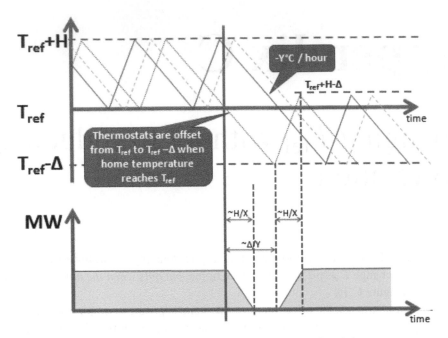

Figure C.1   Decreasing T_{ref} without creating a synchronizing event.

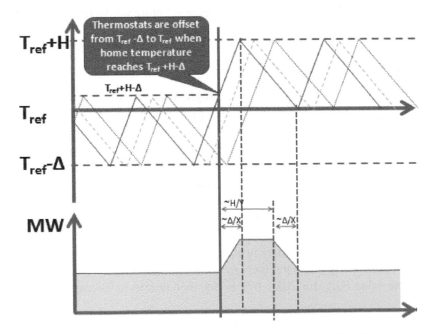

Figure C.2   Adjusting the thermostat back to its normal settings.

Appendix D

Lower Consumption, A Side Benefit of Power Shedding

When the user accepts to live in a home slightly colder than T_{ref}, the thermal dissipation of his home also decreases. The thermal dissipation of the home was $K \times (T_{\text{ref}} - T_{\text{ext}})$ and becomes $K \times (T_{\text{ref}} - T_{\text{ext}} - \Delta)$.

The effect of this lower thermal dissipation is to slightly increase X as the temperature decreases, and to slightly decrease Y (Figure D.1). These effects have been ignored in the evaluations of Appendices B and C.

However, the overall power savings resulting from the lower thermal dissipation during power shedding are easy to evaluate, in a simplified model where we ignore the heating hysteresis H:

- When the temperature is stabilized to $T_{\text{ref}} - \Delta$, the energy used to reheat the houses can be expressed as $P' = P_{\text{n}} \times (T_{\text{ref}} - T_{\text{ext}} - \Delta)/(T_{\text{ref}} - T_{\text{ext}})$: if $T_{\text{ref}} - T_{\text{ext}} = 10\,°C$ and $\Delta = 1\,°C$, P is 10% lower than P'.
- During the temperature decrease period, the dissipation power decreases quasilinearly (we approximate the exponential evolution of the home temperature by a linear function) from P_{n} to P'. If C is the thermal capacity of the house, k the thermal dissipation, t the duration of the house cooling period (heating off) if we neglect the variation of T_{ref}, t' the duration of the house cooling period (heating off) if we take the variation of T_{ref} in consideration we can write:

$$C \times \Delta - k(T_{\text{ref}} - T_{\text{ext}})t = 0$$

$$C \times \Delta - k \int (T_{\text{home}} - T_{\text{ext}})\mathrm{d}t' = 0 \leftrightarrow C \times \Delta - k(T_{\text{ref}} - T_{\text{ext}} - \Delta/2)t'$$

$$= 0 \leftrightarrow t' = t \times (T_{\text{ref}} - T_{\text{ext}})/(T_{\text{ref}} - T_{\text{ext}} - \Delta/2)$$

The Internet of Things: Key Applications and Protocols, First Edition.
Olivier Hersent, David Boswarthick and Omar Elloumi.
© 2012 John Wiley & Sons, Ltd. Published 2012 by John Wiley & Sons, Ltd.

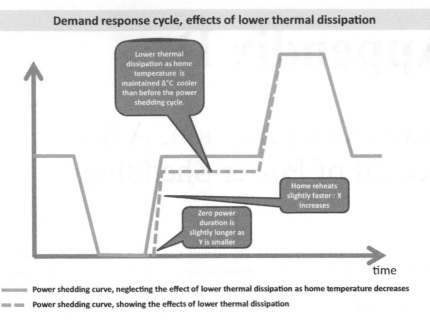

Figure D.1   Effects of the lower thermal dissipation on the power shedding cycle.

Therefore, with the same values of $(T_{ref} - T_{ext})$ and Δ as above, we can estimate the real house cooling period with the heating off to be about 5% longer than in our estimates of Appendices B and C.

– During the temperature increase period, the dissipation power increases linearly from P' to P_n. If P_{max} is the power of the heating system we can write:
  • $P_{max}t - k(T_{ref} - T_{ext})t = C \times \Delta$ (energy conservation as the house reheats).
  • Mean heating power $= P = P_{max}Y/(X + Y) = k(T_{ref} - T_{out})$.
  • $-C \times \Delta + P_{max}t' - k \int (T_{home} - T_{ext})dt' = 0 \leftrightarrow -C^*\Delta + P_{max}t' - k(T_{ref} - T_{ext} - \Delta/2)t' = 0$.
  • By combining the above we obtain $t' = t[(T_{ref} - T_{ext})X/Y]/[(T_{ref} - T_{ext})X/Y + \Delta/2]$.
  • With the same values of $T_{ref} - T_{ext}$ and Δ as above, $X = 1{,}6\,°/h$ and $Y = 0{,}6\,°/h$ we see that the real heating time is about 2% shorter than what we obtain in Appendices B and C, when taking into account the variation of the house thermal dissipation during the reheat cycle. We obviously have the same ratio regarding the reheat energy, we use $t'P_{max}$ instead of tP_{max}.

In summary, the impact of the variation of the house thermal dissipation during the reheat cycle can be summarized as follows:

- The evaluations of Appendices B and C for the energy made available to the grid during the power shedding cycle are underevaluated by about 5%. The undervaluation becomes more important as $T_{ref} - T_{ext}$ decreases (mild temperatures).
- The impact on the reheat cycle is negligible.
- The whole cooling/reheat cycle is not completely neutral for the home owner. He gains about 5% of the energy made available to the network as a result of the imbalance between the energy savings of the cooling period and the extra energy used during the reheat cycle. In addition, his bill is reduced by about 10% as long as the temperature is maintained 1 °C colder.

Printed and bound by CPI Group (UK) Ltd, Croydon, CR0 4YY

27/10/2024

14580199-0005

Index

The Internet of Things: Key Applications and Protocols, First Edition.
Olivier Hersent, David Boswarthick and Omar Elloumi.
© 2012 John Wiley & Sons, Ltd. Published 2012 by John Wiley & Sons, Ltd.

Printed and bound by CPI Group (UK) Ltd, Croydon, CR0 4YY